Autodesk Fusion 360 CAM Overview

Tutorial Books

Download Resource Files from:

online.books999@gmail.com

Contents

TUTORIAL 1

This tutorial takes you through the creation of your first machining operation in Autodesk Fusion 360.

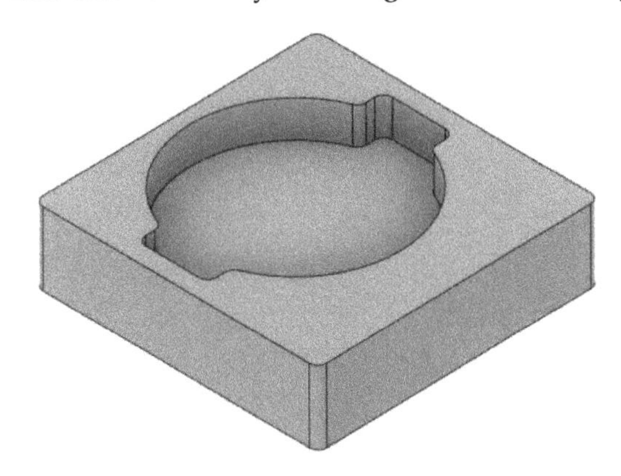

Creating a New Project

1. Start **Autodesk Fusion 360** by double-clicking the **Autodesk Fusion 360** icon on your desktop.
2. To create a new project, click the **Show Data Panel** button located at the top left corner of the window.
3. Click the **Return to project list** button.
4. Click the **New Project** button.

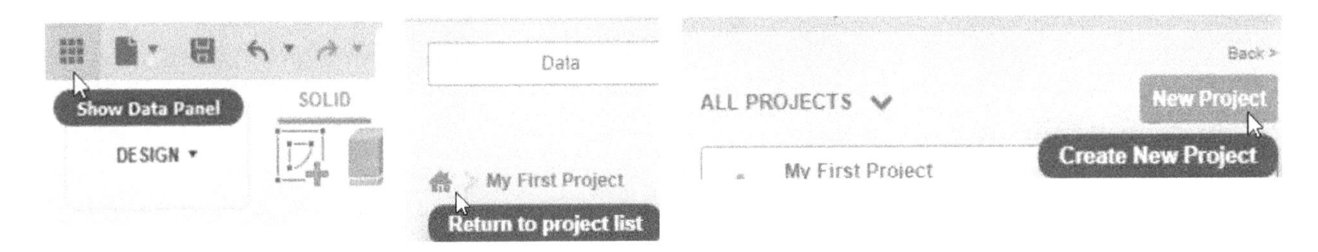

5. Type **Fusion 360 CAM Overview** and press Enter. Next, click the **Close Data Panel** button.
6. On the top-right corner, click **Help > Quick Setup**.

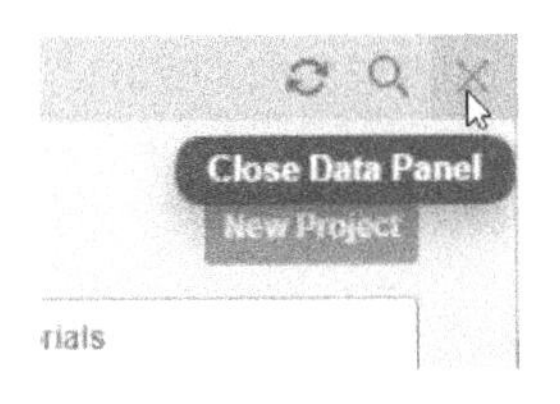

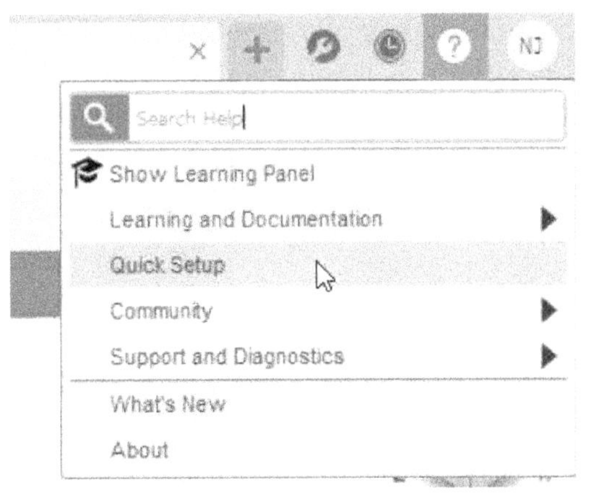

7. On the **Quick Setup** dialog, select **Default Units > in**.
8. Select **CAD Experience > New to CAD**.
9. Click **Close** on the **Quick Setup** dialog.

Uploading the CAD Model

1. Download the **Tutorial 1** file.
2. On the Application Bar, click **File** drop-down > **Upload**.
3. Click the **Change Location** option next to the **Location** box.
4. Select the **Fusion 360 CAM Overview** project from the Project list.
5. Click the **Select** button. Next, click the **Select Files** button.
6. Browse to the location of the downloaded file. Next, double-click on the Tutorial 1 file.
7. Click the **Upload** button. Next, close the **Job Status** dialog.
8. Click the **Show Data Panel** button located at the top left corner of the window.
9. Double-click on the **Fusion 360 CAM Overview** project to open it.
10. Double-click on the **Tutorial 1** file to open it.

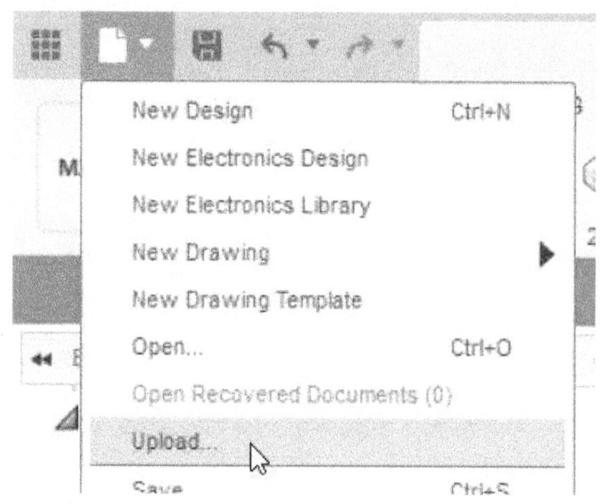

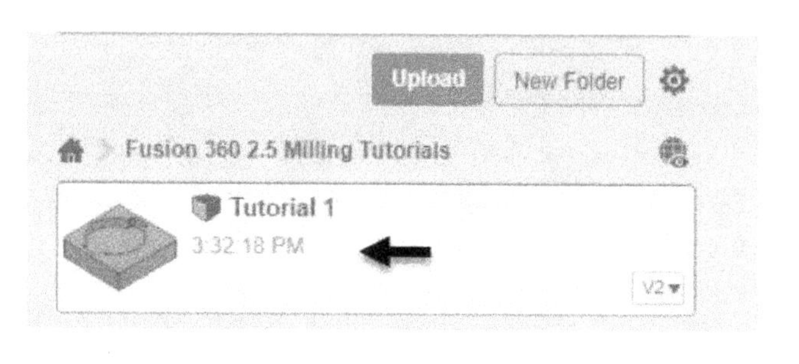

11. Close the **Data** panel.

Creating a new Manufacturing Model

In this section, you will create a manufacturing model from the design. By doing so, you can edit the manufacturing model without changing the original design. However, the changes made to the original design will carry forward to the manufacturing model. Also, you can add a fixture or vise to the manufacturing model.

1. On the Toolbar, click **Change Workspace** drop-down **> Manufacture**.
2. In the Browser, place the pointer on the **Units** node. Next, click **Change Active Units**.

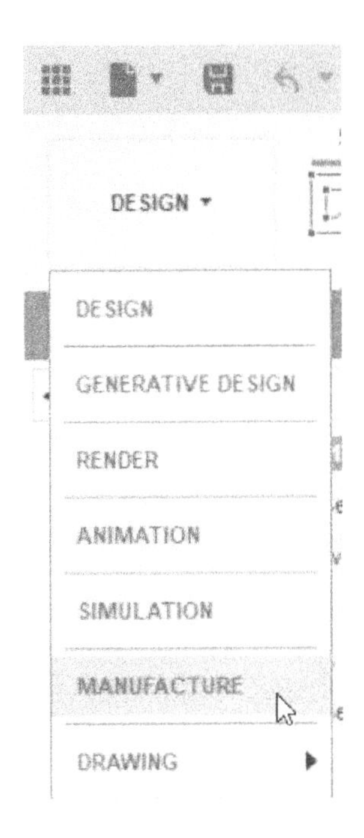

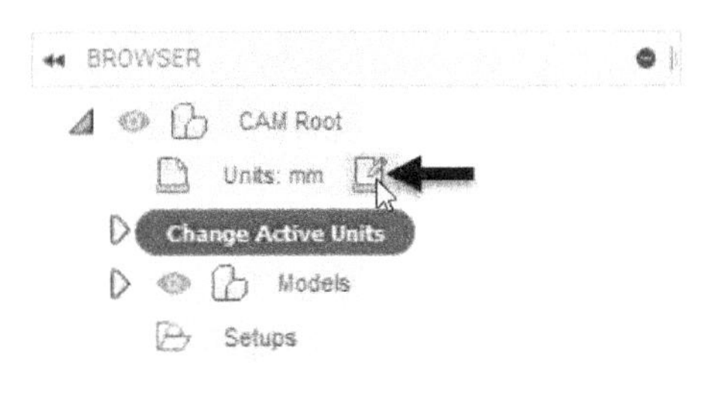

3. On the **Change Active Units** dialog, select **Unit Type > Inch**. Next, click **OK**.
4. On the Toolbar, click **Milling > Setup > Create Manufacturing Model.** Notice that the **Manufacturing Model** node is added to the Browser. You can edit the manufacturing model by double-clicking it (or) click the right mouse button on the **Manufacturing Model** node and select **Edit Manufacturing Model**.

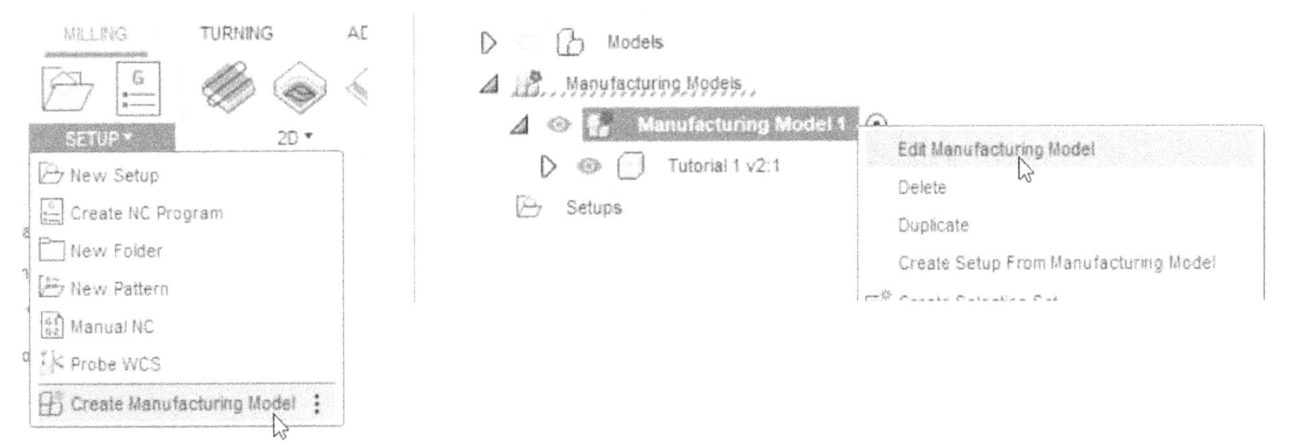

Notice that there are three additional editing tools on the Toolbar. Use the tools to edit the model and click the **Finish Edit** button.

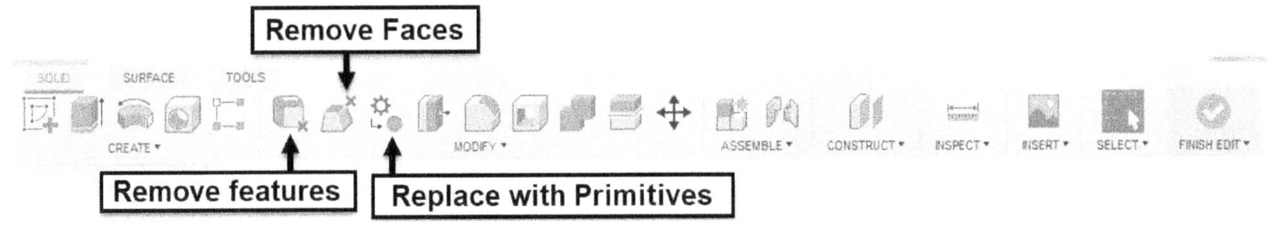

Creating a new Manufacturing Setup

In this section, you will create a new manufacturing setup:

1. On the Toolbar, click **Milling > Setup > Setup.**

Notice the transparent yellow box (stock) enclosing the CAD model. Also, the work coordinate system is displayed on the yellow box.

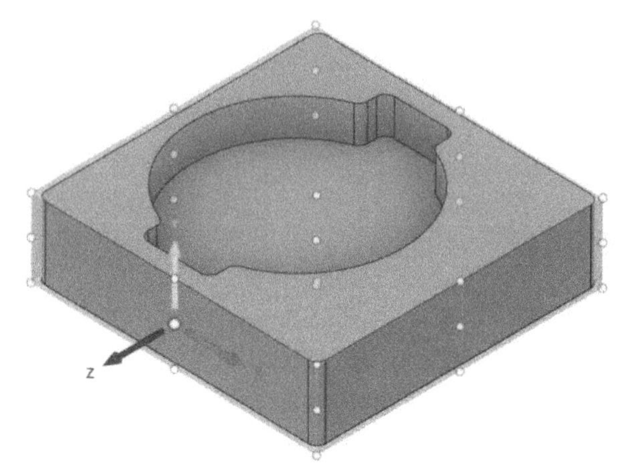

2. On the **Setup** dialog, click the **Select** button next to the **Machine** option.
3. On the **Machine Library** dialog, select the **Fusion 360 Library** option from the left side.
4. Select the **Autodesk Generic 3-axis** machine from the list.

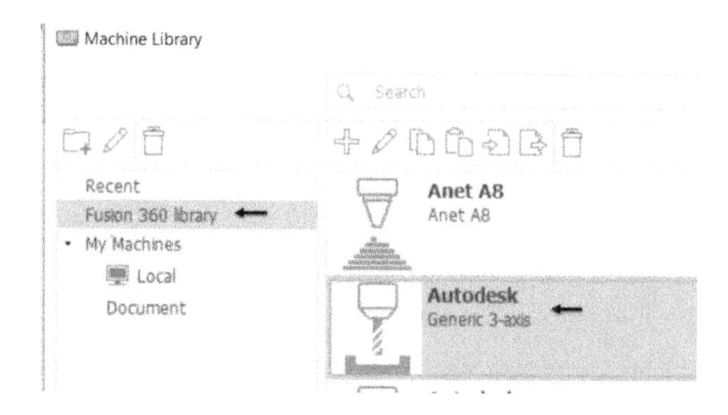

5. Click the **Select** button.
6. On the **Setup** dialog, select **Operation Type > Milling**.
7. On the **Setup** dialog, select **Orientation > Select Z Axis/plane & X axis.**
8. Select the vertical edge to define the Z axis, as shown.
9. Select the front horizontal edge to define the X axis.

Ensure that the Z-axis, X-axis, and Y-axis point upward, right, and backward, respectively.

10. Check the **Flip Z Axis** option if the Z axis is pointing downward.
11. Click the **Box point** button next to the **Stock Point** option and select the stock to locate at the top face's center.

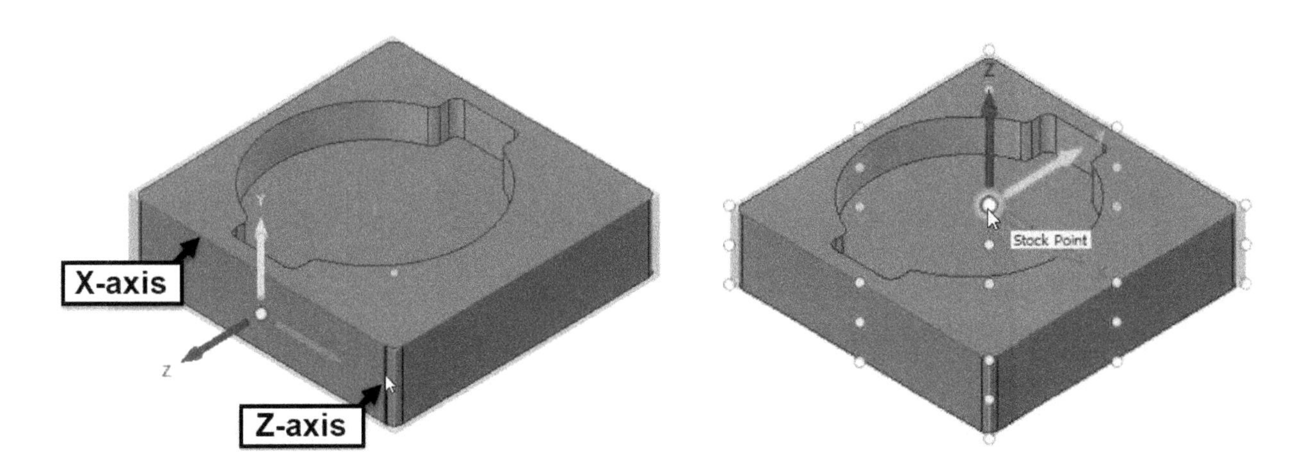

12. Under the **Model** section, make sure that the **Manufacturing Model** is set to **Manufacturing Model 1**.
13. Click the **Stock** tab on the **Setup** dialog.
14. Select **Mode > Fixed size box**. Next, type **4.1** in the **Width (X)** and **Depth (Y)** boxes, respectively.
15. Type **1.35** in the **Height (Z)** box.
16. Set the **Model Position** to **Center** along the **Width (X)** and **Depth (Y)** directions.

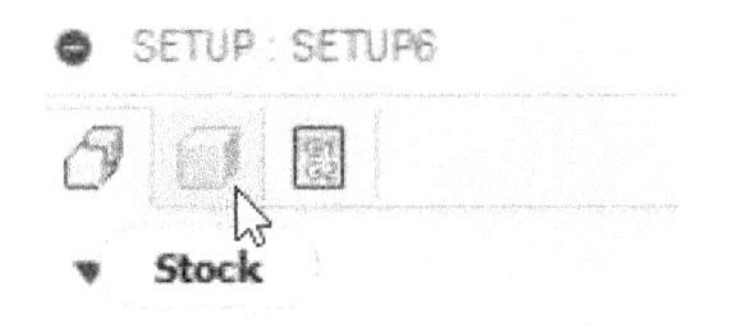

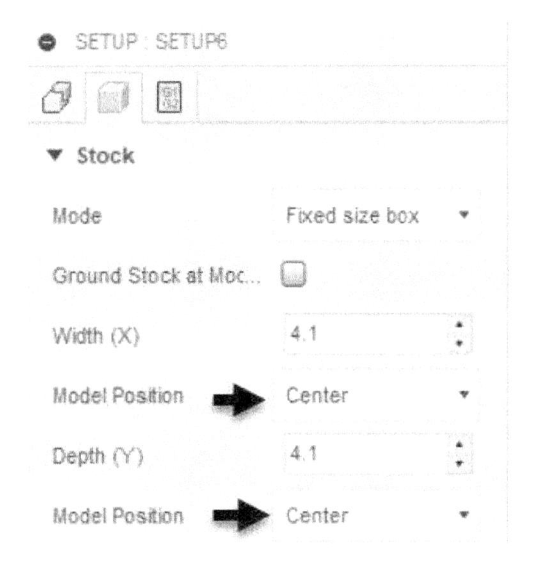

17. Set the **Model Position** to **Offset from top (+Z)** along the **Height (Z)** direction.
18. Enter **0.01** in the **Offset** box. Next, click **OK** on the **Setup** dialog.
19. Select the Front face of the ViewCube.

Notice the extra material at the bottom of the workpiece. It helps to hold the workpiece.

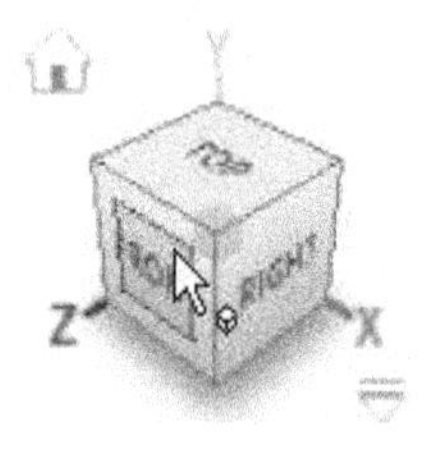

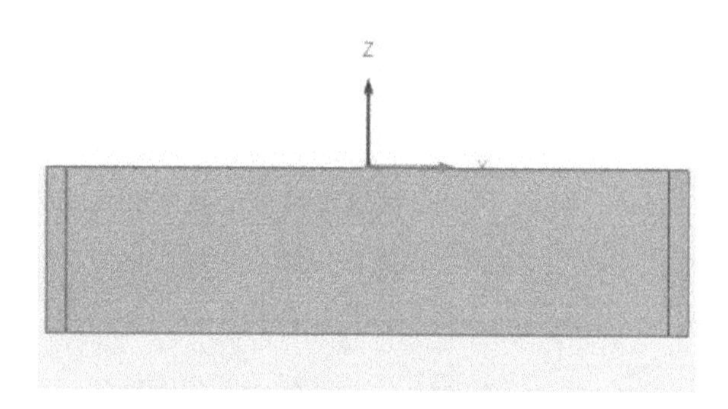

20. Click in the graphics window and notice that the stock disappears.
21. Click the **Home** icon next to the ViewCube to change the view orientation to Isometric.
22. Select the **Setup** under the Browser to view the stock.

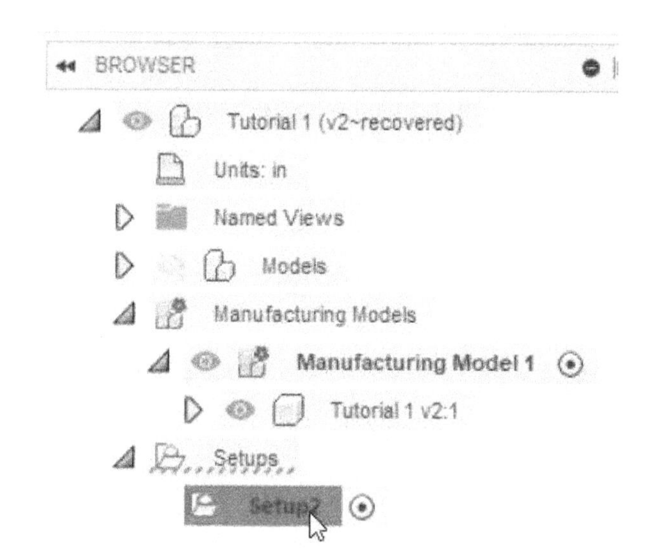

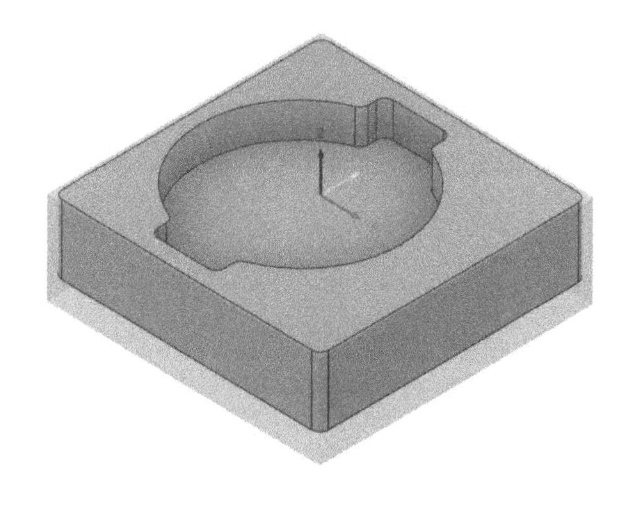

Creating the Facing Operation

Facing is the first operation during the milling process.

1. On the Toolbar, click **Milling > 2D > Face.**
2. On the **Face** dialog, click the **Tool** tab and select the **Tool** button; the **Select Tool** window pops up on the screen.

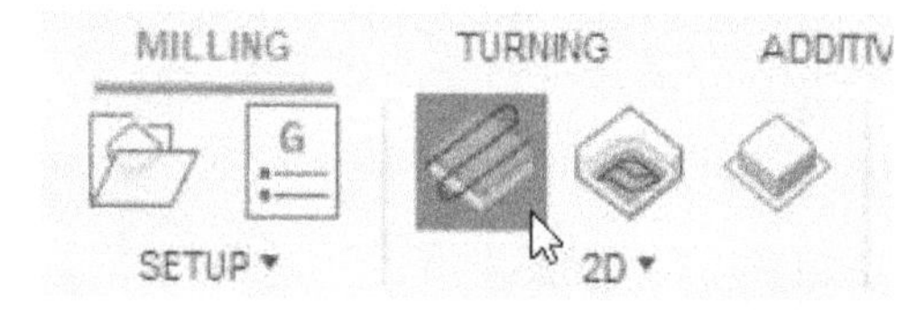

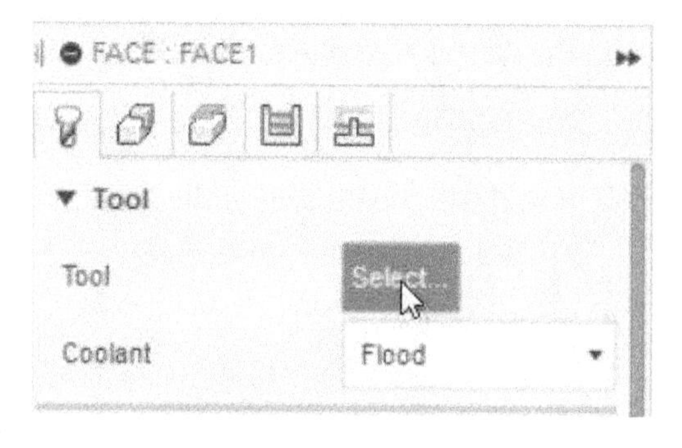

The **Select Tool** window displays all the tool libraries on the left side.

3. On the **Select Tool** window, select the **Tutorial – Inch** folder from the **Fusion 360 Library**.
4. Select the **1-Ø2" L1.563" (2" Face Mill)** tool from the tool list.

5. Click the **Select** button located at the bottom-right corner.

The preview of the selected tool is displayed in the graphics window. Also, the feed and speed values are populated in the **Feed & Speed** section of the **Face** dialog. These are the default values assigned to the tool when it was created in the library.

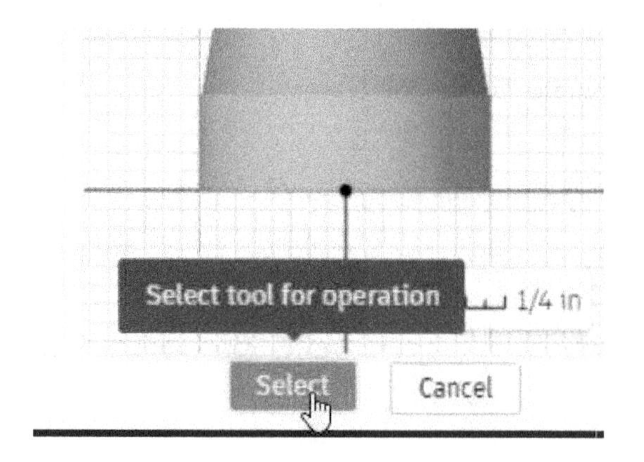

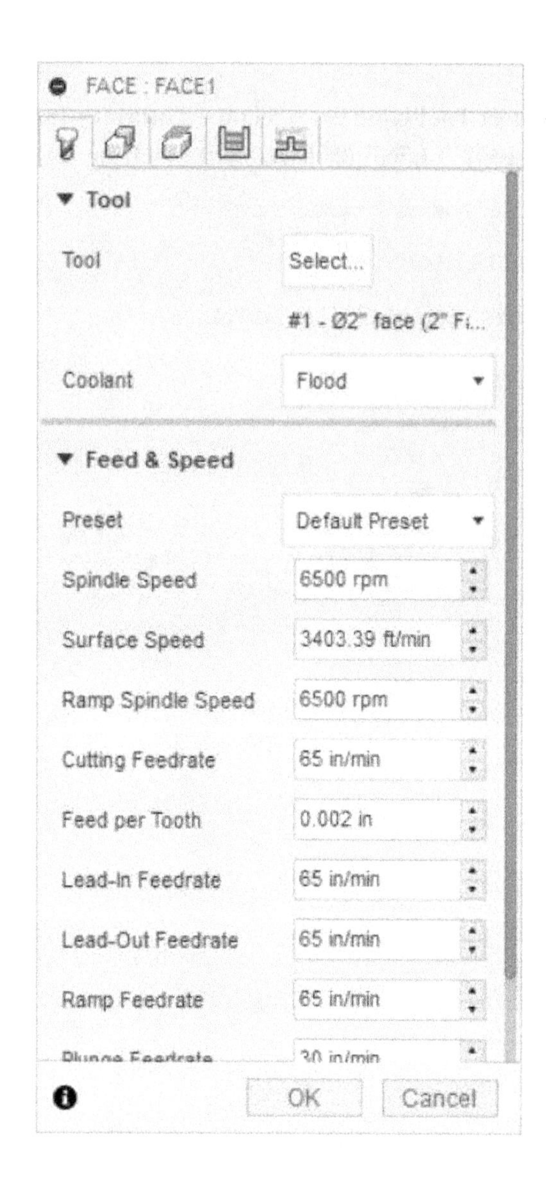

6. Click the **Geometry** tab on the **Face** dialog.

Now, you need to select a face. However, you can leave this option unselected to face the complete stock.

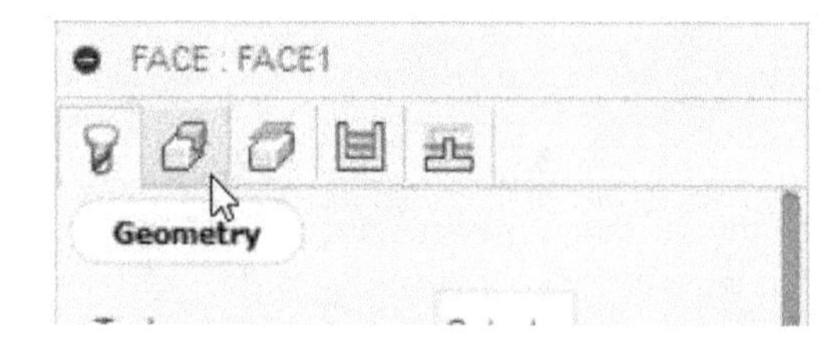

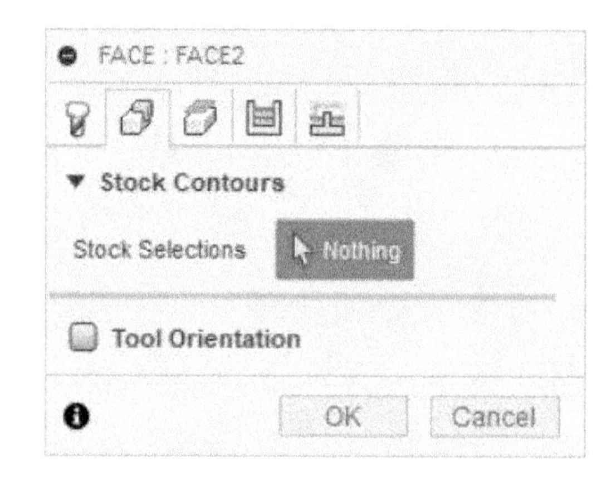

7. Click the **Heights** tab on the **Face** dialog.

On this tab, you need to define the **Top Height** and the **Bottom Height**. The top height defines the starting height of the cutter from where the material will be removed. The bottom height defines the depth up to which the material will be removed.

8. In the **Top Height** section, select **From > Stock top**. The tool starts cutting from the top face of the stock.
9. Leave the **Offset** value to **0**.
10. In the **Bottom Height** section, select **From > Model top**. The tool removes the material up to the top face of the model.
11. Leave the **Offset** value to **0**.

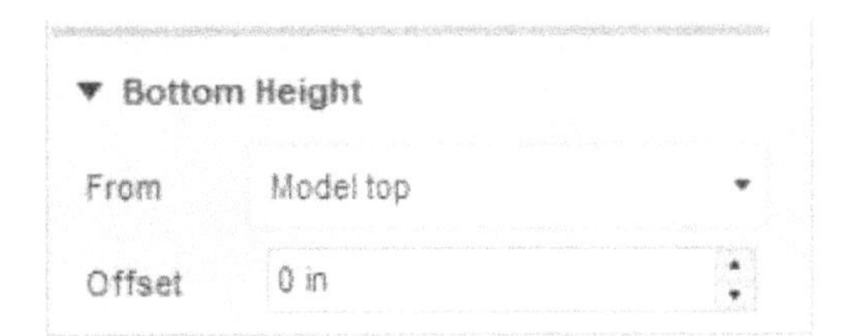

12. Click the **Passes** tab on the **Face** dialog. Next, type **0.25** in the **Stock Offset** box.

The **Stock Offset** value defines the distance up to which the facing tool moves beyond the stock. It helps you to machine the complete stock without leaving the edges.

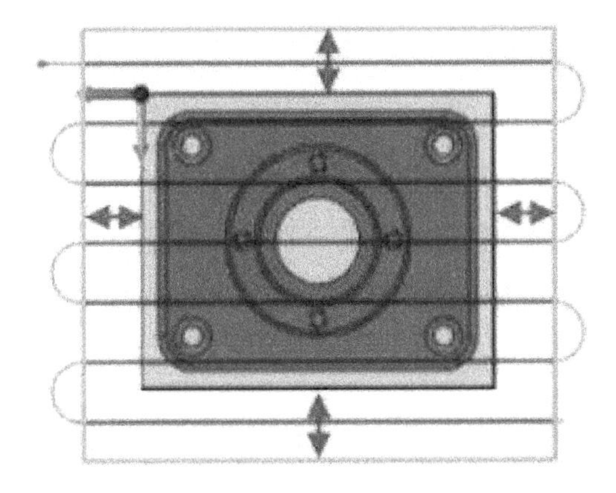

Stock Offset

13. Leave the default value in the **Stepover** box. The default **Stepover** value is 95% of the tool diameter. The stepover value is the distance between the two passes.

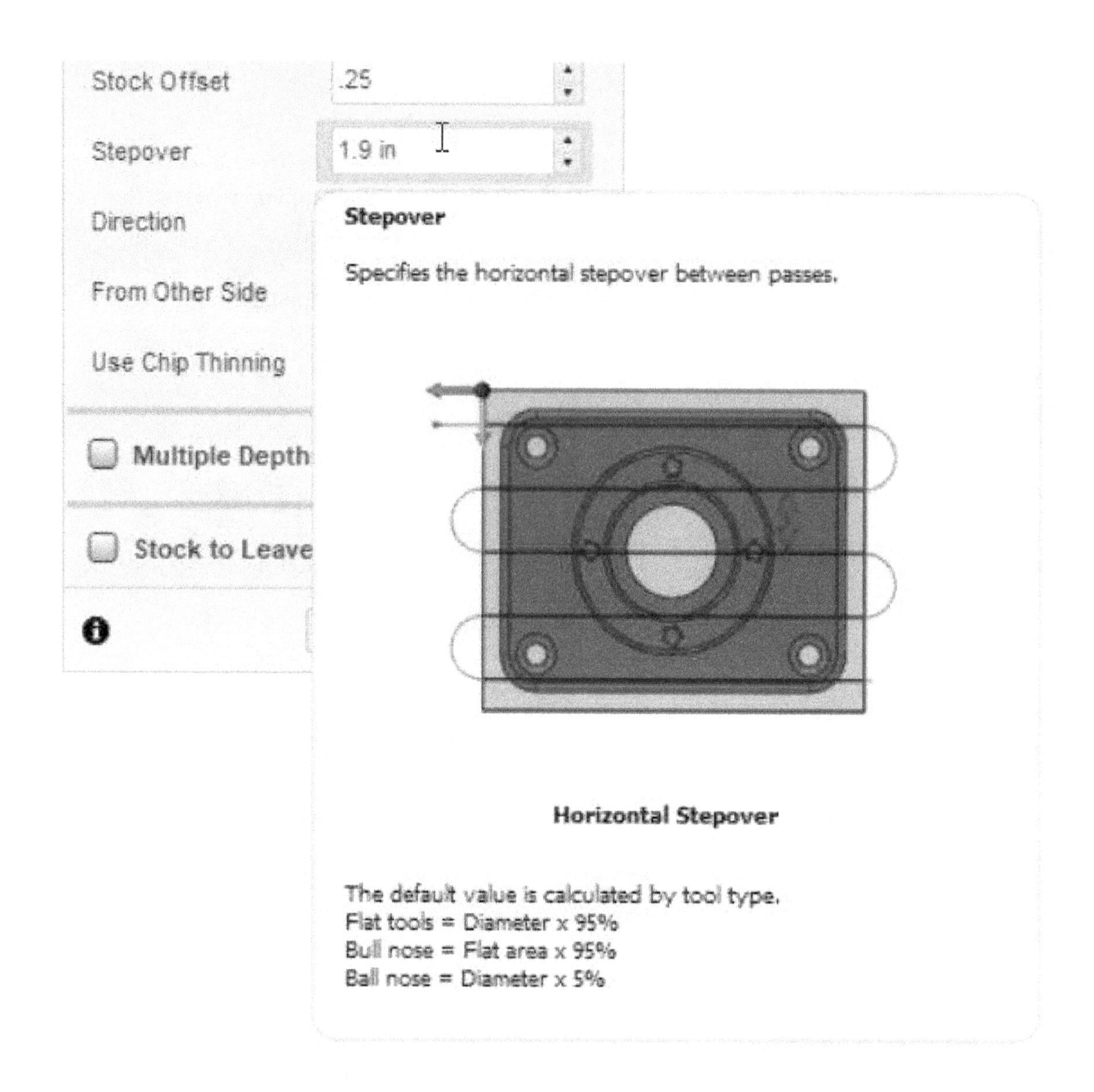

The **Stepover** value is calculated based on an expression. However, you can edit the expression by right-clicking in the **Stepover** box and selecting **Edit Expression**; the **Expression** dialog appears. Change the expression on the **Expression** dialog and click **OK**.

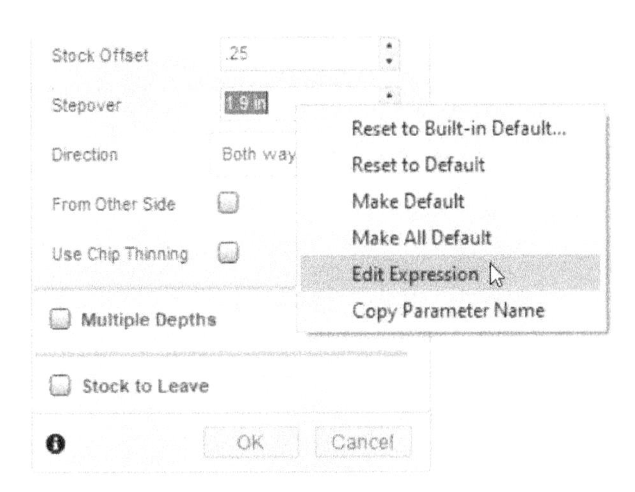

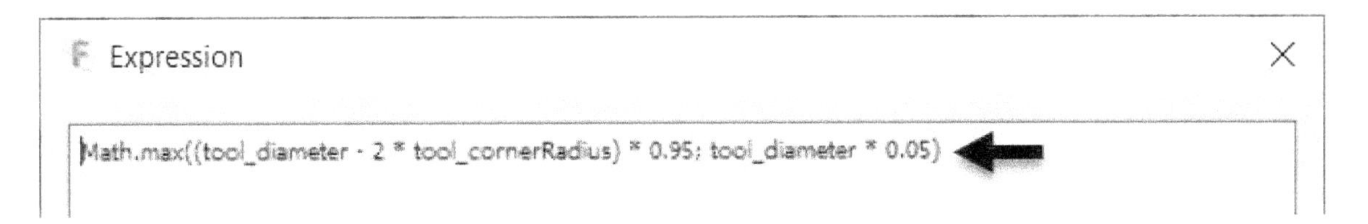

14. Click the **Linking** tab on the **Face** dialog. The options in this tab define the way the tool enters and exits the material. Also, the transitions between the toolpaths are defined.
15. Check the **Extend Before Retract** option. It extends the cutting pass beyond the stock by half of the tool diameter before retracting.
16. Click **OK** on the **Face** dialog to create the face milling operation.

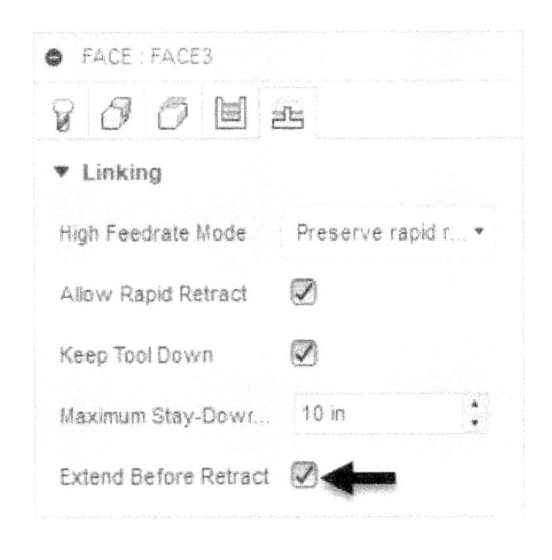

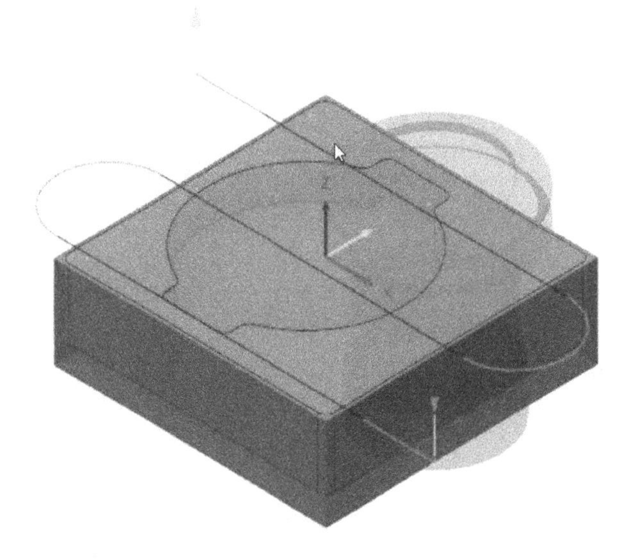

The toolpath has three colors (yellow, green, and blue). The yellow color indicates the Linking move, which means that the tool moves IN and OUT of the cut without removing any material.

The green color indicates the LEAD IN and LEAD OUT of the toolpath.

The blue color indicates the actual cut.

Simulating the Facing Operation

1. Select the **Face** operation from the Browser.
2. On the Toolbar, click **Milling > Actions > Simulate**.

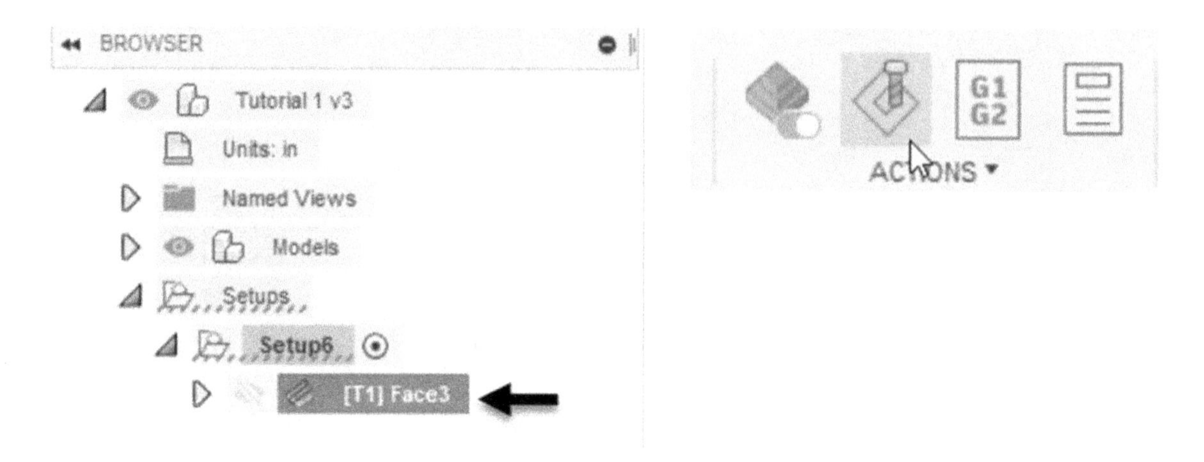

The options to play the simulation are displayed at the bottom of the screen.

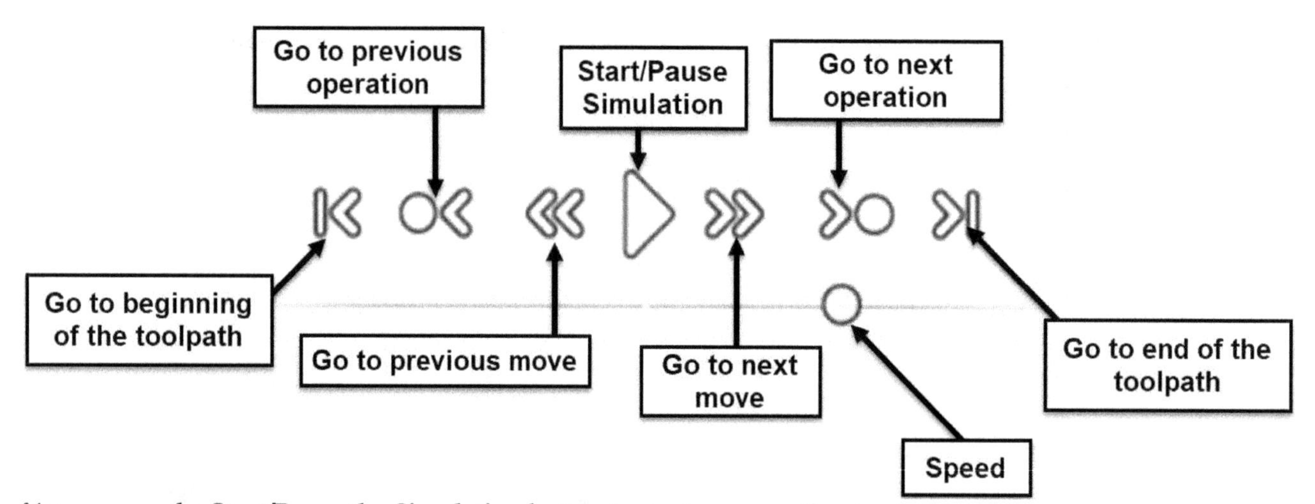

You can use the **Start/Pause the Simulation** button to start or pause the simulation.

You can adjust the simulation speed using the simulation dragger displayed below the icons.

Place the pointer on the remaining buttons to know their functions.

On the **Simulate** dialog, keep the **Tool** option checked to display the tool. You can use the **Tool** drop-down to display the **Holder**, **Flute**, or **Shaft** of the cutting tool. You can also turn ON/OFF the transparency using the **Transparent** check box.

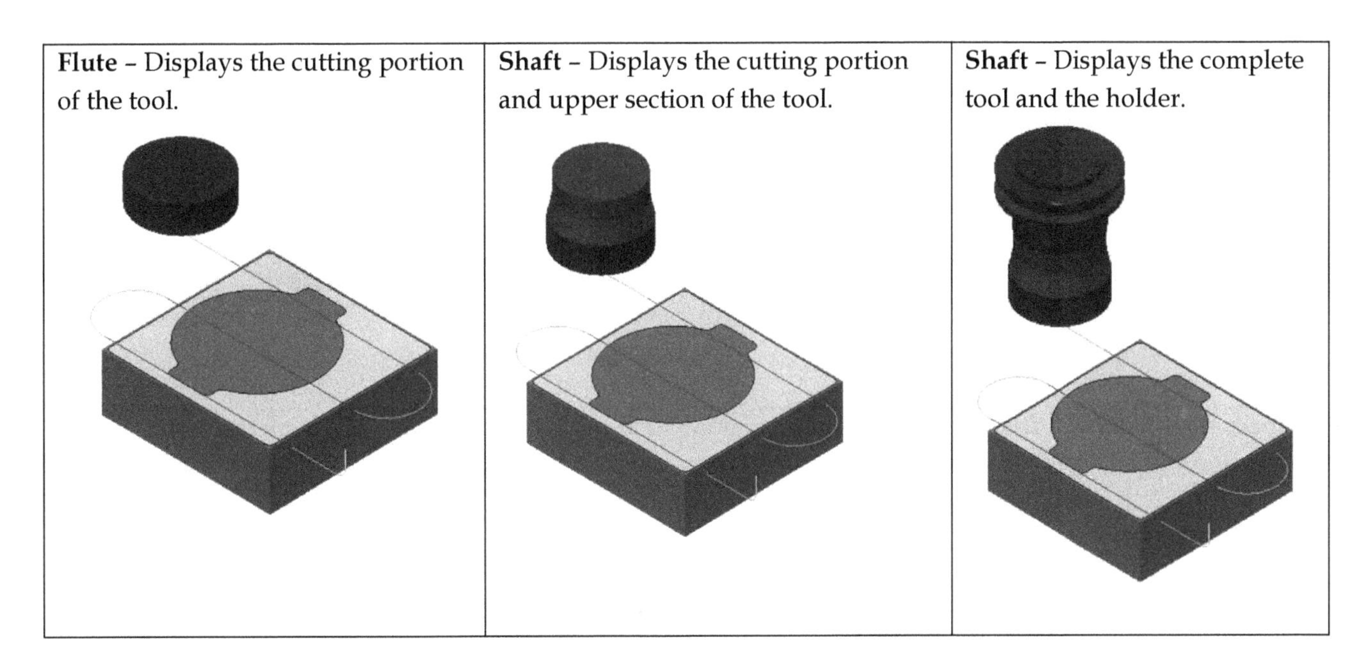

Flute - Displays the cutting portion of the tool.	Shaft - Displays the cutting portion and upper section of the tool.	Shaft - Displays the complete tool and the holder.

You can check/uncheck the **Toolpath** option to display or hide the toolpath. Next, use the **Mode** drop-down to specify the way the toolpath is displayed.

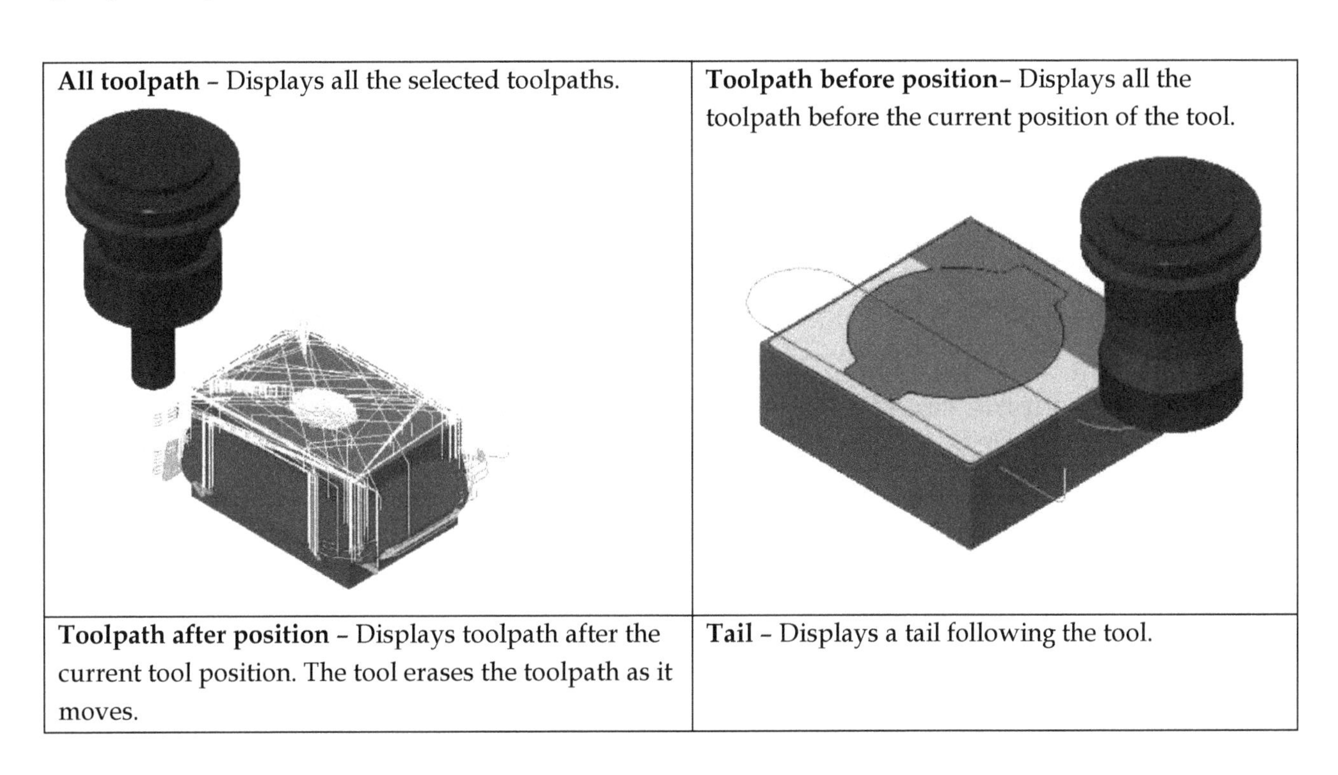

All toolpath - Displays all the selected toolpaths.	Toolpath before position- Displays all the toolpath before the current position of the tool.
Toolpath after position - Displays toolpath after the current tool position. The tool erases the toolpath as it moves.	Tail - Displays a tail following the tool.

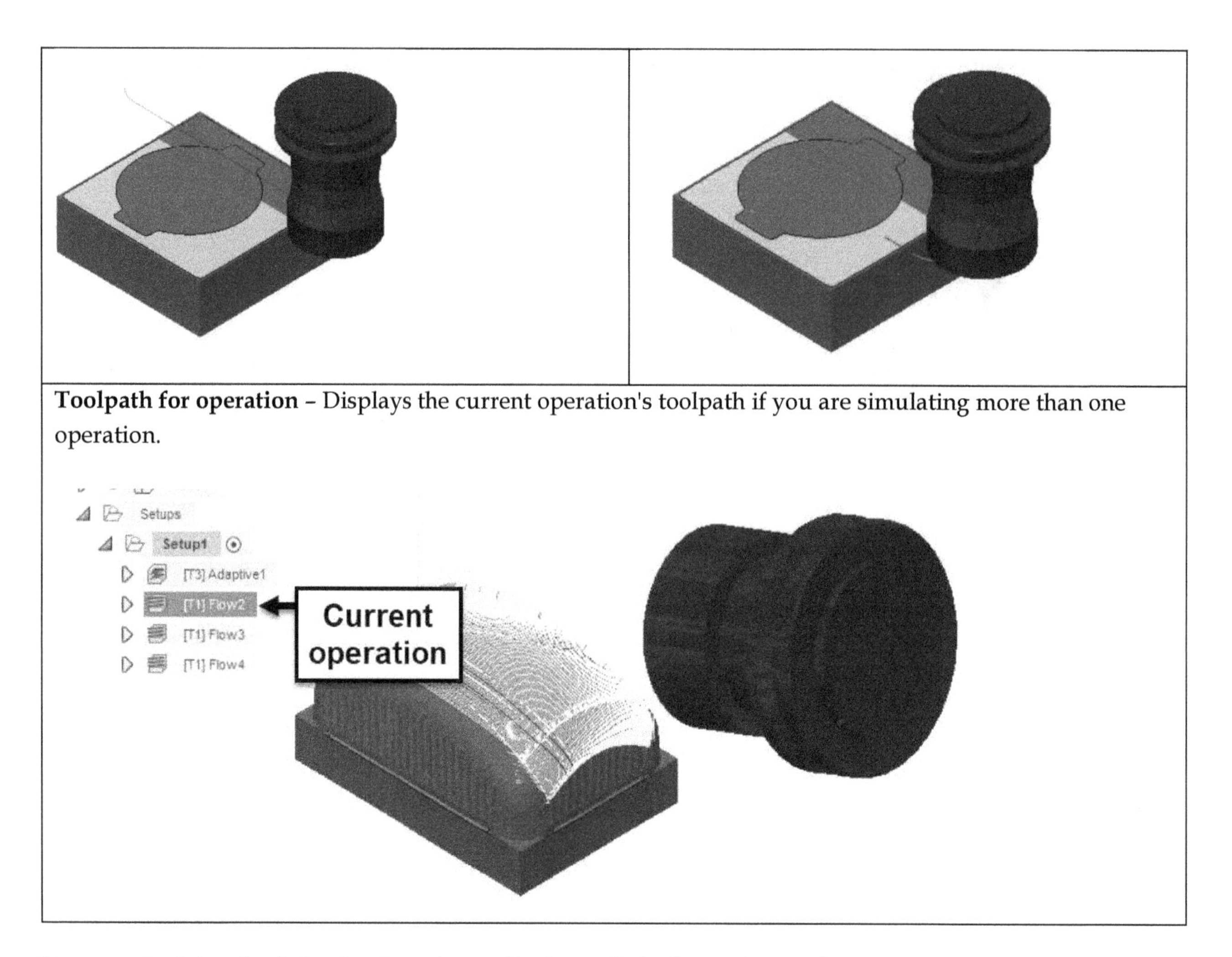

Toolpath for operation – Displays the current operation's toolpath if you are simulating more than one operation.

You can check/uncheck the **Stock** option to display or hide the stock. Use the **Mode** drop-down to specify the stock mode.

The **Standard** option displays the stock for all the toolpaths (from 2-axis to 5-axis toolpaths). Also, it helps you to detect the undercuts and collisions.

The **Fast for 3X only** option is used for faster performance. However, it displays the stock only up to the 3-axis toolpaths without any collision detection. The advantage of this option is that it offers additional stock colorization.

Use the **Colorization** drop-down to specify the stock colorization.

Material – This option displays the stock in the material selected from the **Material** drop-down.

Operation – This option displays the stock in color based on the operation.

Tool – This option displays the stock in the color based on the tool.

Comparison – This option displays the excess stock material in blue, gouged material in red, and in-tolerance material in green.

Use the **Material** drop-down to specify the stock material.

3. Under the Stock section, select **Mode > Fast for 3X only**.
4. Select **Quality > High** on the **Simulate** dialog.

Click the **Info** tab on the **Simulate** dialog to view the tool's position, spindle speeds, feed rate, and movement.

The **Operation** section displays the operation's information, such as type, tool, setup, work offset, and time.

The **Machine** section displays the machine used for the operation.

The **Verification** section displays information such as collisions, volume, start volume, and distance. Click the **Continue** button to view the information.

Click the **Statistics** tab on the **Simulate** dialog to view the machining time, machining distance, operations, and tool changes.

5. Click the **Start the Simulation** button located at the bottom of the screen.

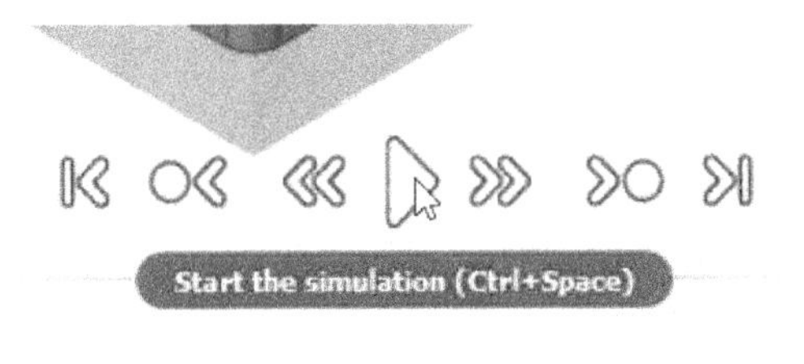

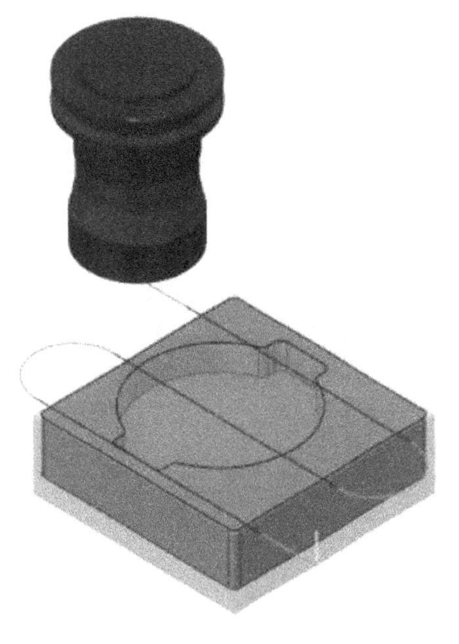

6. Click the **Close** button on the **Simulate** dialog.

Creating the 2D Adaptive Clearing Operation

The 2D Adaptive Clearing operation is a roughing operation. It is used to remove large amount of material before performing a finishing operation.

1. On the Toolbar, click **Milling > 2D > 2D Adaptive Clearing**.
2. On the **2D Adaptive** dialog, click the **Select** button next to the **Tool** option.
3. On the **Select Tool** window, select the **Tutorial - Inch** folder from the **Fusion 360 Library**.
4. Select the **4-Ø1/4" L1" (flat end Mill)** tool from the tool list.
5. Click the **Select** button located at the bottom-right corner.

6. Click the **Geometry** tab on the **2D Adaptive** dialog.
7. Click on the floor face of the pocket, as shown. The selected face is turned blue, which indicates the region to be cut.

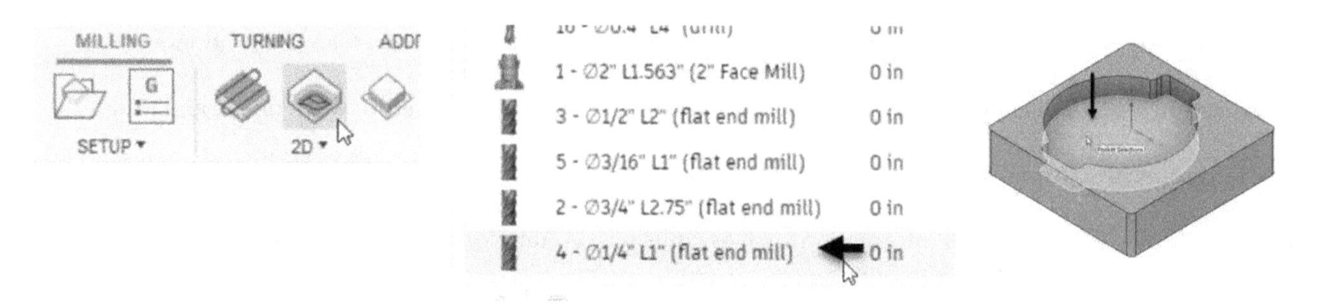

8. Click the **Heights** tab on the **2D Adaptive** dialog.
9. In the **Top Height** section, select **From > Model top**. The tool starts cutting from the top face of the model.
10. Leave the **Offset** value to **0**.
11. In the **Bottom Height** section, select **From > Selected contours(s)**. The tool removes the material up to the selected face.
12. Leave the **Offset** value to **0**.
13. Click the **Passes** tab on the **2D Adaptive** dialog.

Now, you need to define the **Optimal load** value. The **Optimal load** value is the maximum diameter of the tool that engages with the workpiece. By default, Autodesk Fusion 360 calculates this value as 40% of the tool diameter. However, you can define your value using an expression.

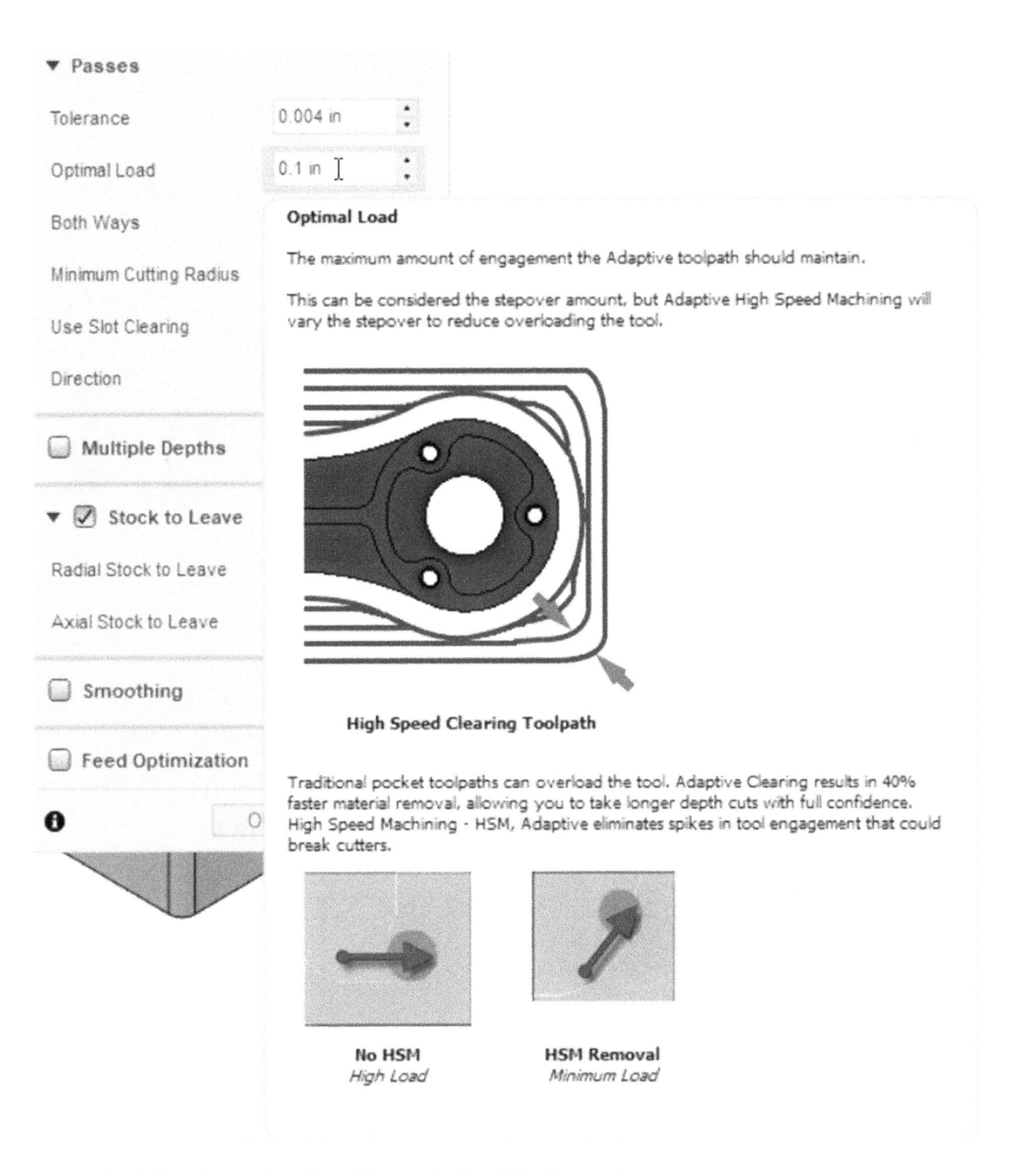

14. Right-click in the **Optimal Load** box and select **Edit Expression**.
15. Change the expression to **tool_diameter*0.25**.

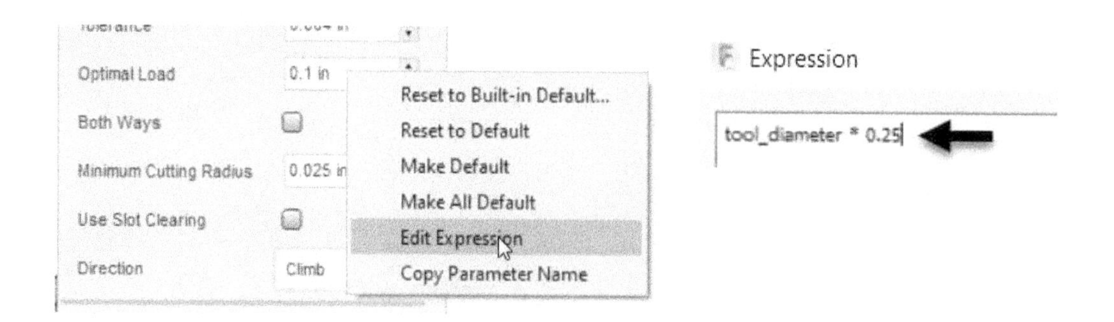

16. Click **OK** on the **Expression** dialog; the **Optimal Load** value is changed to 0.0625.
17. Check the **Multiple Depths** option. The **Multiple Depths** section is expanded. The options in this section are used to specify the maximum depth of the cutting operation. The rule for the maximum depth is 200xTool Diameter. However, the maximum depth should be lower to avoid tool breakage.

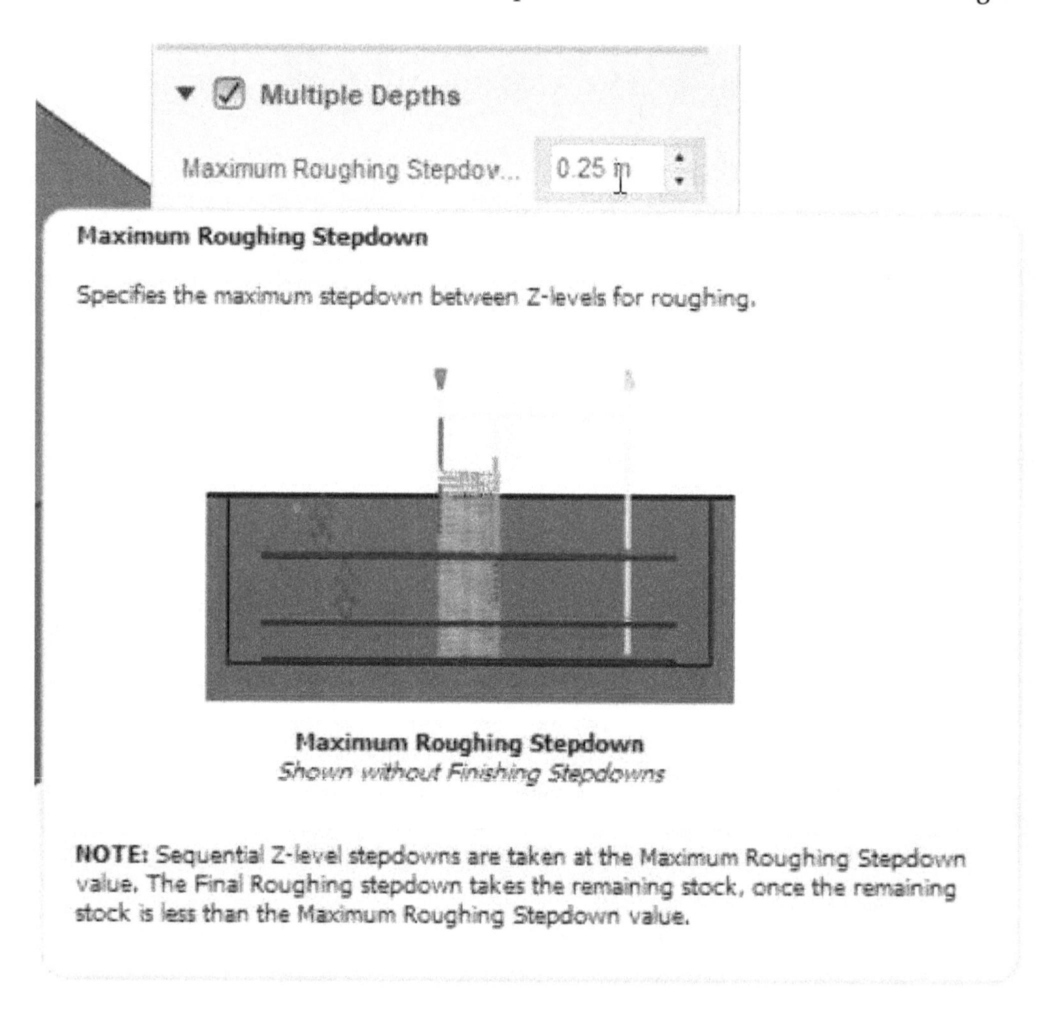

18. Change the **Maximum Roughing Stepdown** value to **0.25**.
19. Change the **Radial Stock to Leave** value to **0.01**.
20. Change the **Axial Stock to Leave** value to **0.00**.
21. Click the **Linking** tab on the **2D Adaptive** dialog.

Notice that the **Ramp Type** and **Ramping Angle** are set to **Helix** and **2 deg**, respectively.

22. Click **OK** on the **2D Adaptive** dialog to create the adaptive clearing milling operation.
23. Select the **2D Adaptive** operation from the Browser.
24. On the Toolbar, click **Milling > Actions > Simulate**.

25. Click the **Start the Simulation** button located at the bottom of the screen. The 2D Adaptive operation is simulated.

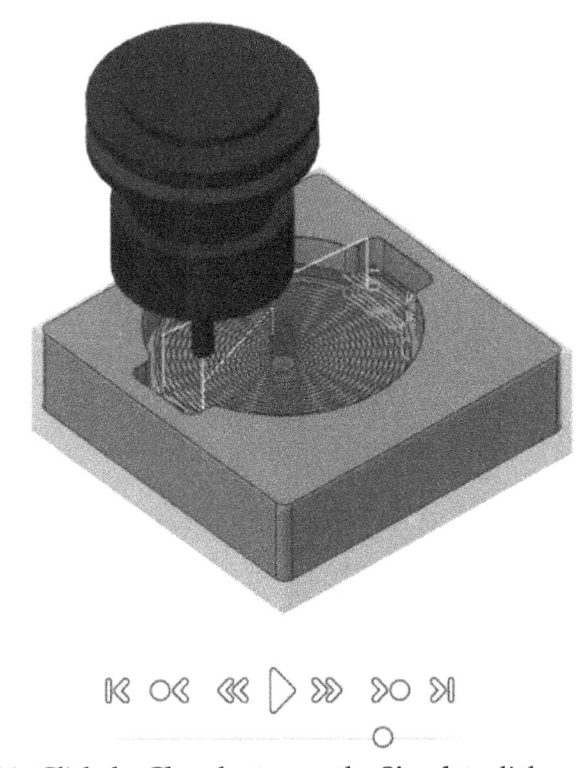

26. Click the **Close** button on the **Simulate** dialog.

Creating the 2D Contour Operation

After creating the roughing operation using the 2D Adaptive Clearing, you need to machine the side faces using the 2D Contour operation.

1. Click **2D Contour** on the **2D** panel.

Notice that the **4-Ø1/4" L1" (flat end Mill)** tool is already selected.

2. Click the **Geometry** tab on the **2D Contour** dialog.
3. Click on the edge of the pocket, as shown.

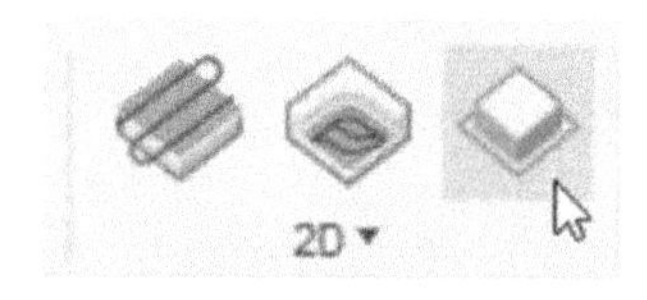

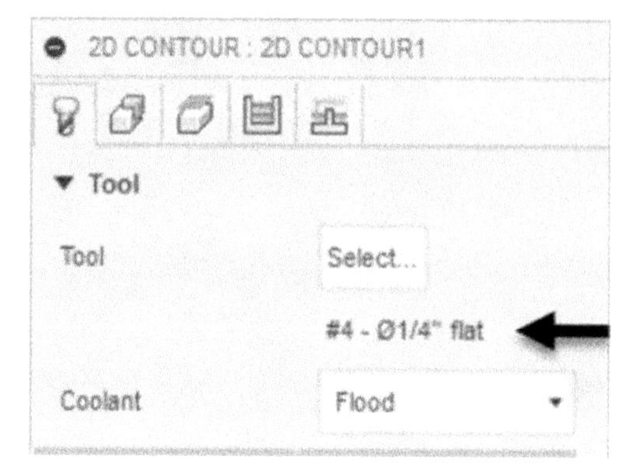

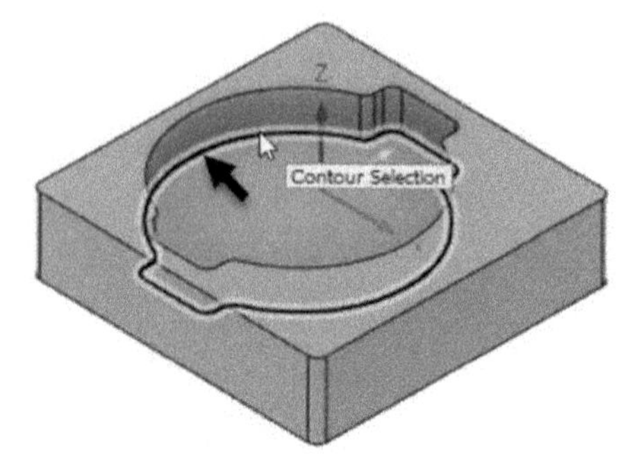

4. Click the **Heights** tab on the **2D Contour** dialog.
5. In the **Top Height** section, select **From > Model top**. The tool starts cutting from the top face of the model.
6. In the **Bottom Height** section, select **From > Selected contours(s)**.
7. Click the **Passes** tab on the **Face** dialog.

Now, you need to define the **Finishing Overlap** value. The **Finishing Overlap** value is the distance the tool travels beyond the entry point.

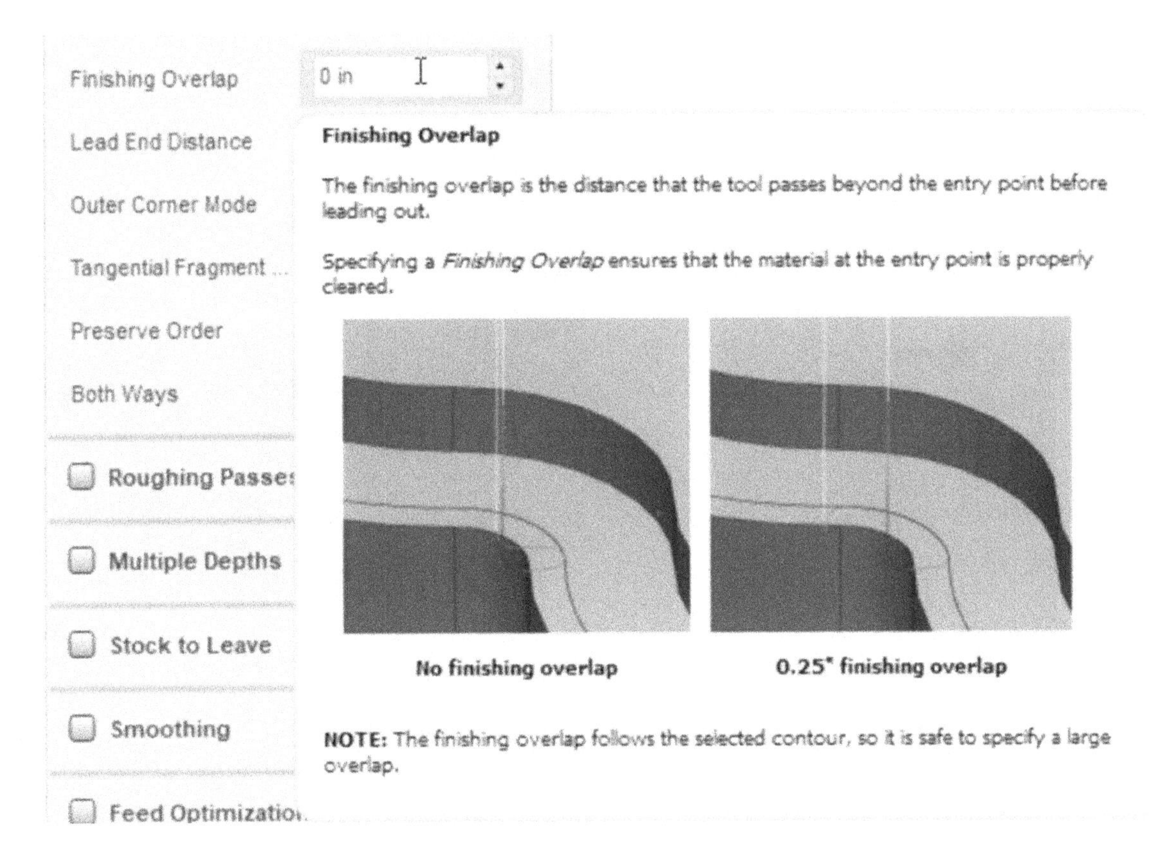

8. Change the **Finishing Overlap** value to 0.05.
9. Leave the default settings on the **Linking** tab.
10. Click **OK** on the **2D Contour** dialog to clean up the wall of the pocket.
11. Select the **Setup** tab from the Browser.
12. On the Toolbar, click **Milling > Actions > Simulate**.
13. On the **Simulate** dialog, make sure that the **Mode** is set to **Toolpath before position**.
14. Click the **Start the Simulation** button located at the bottom of the screen. All three operations are simulated.

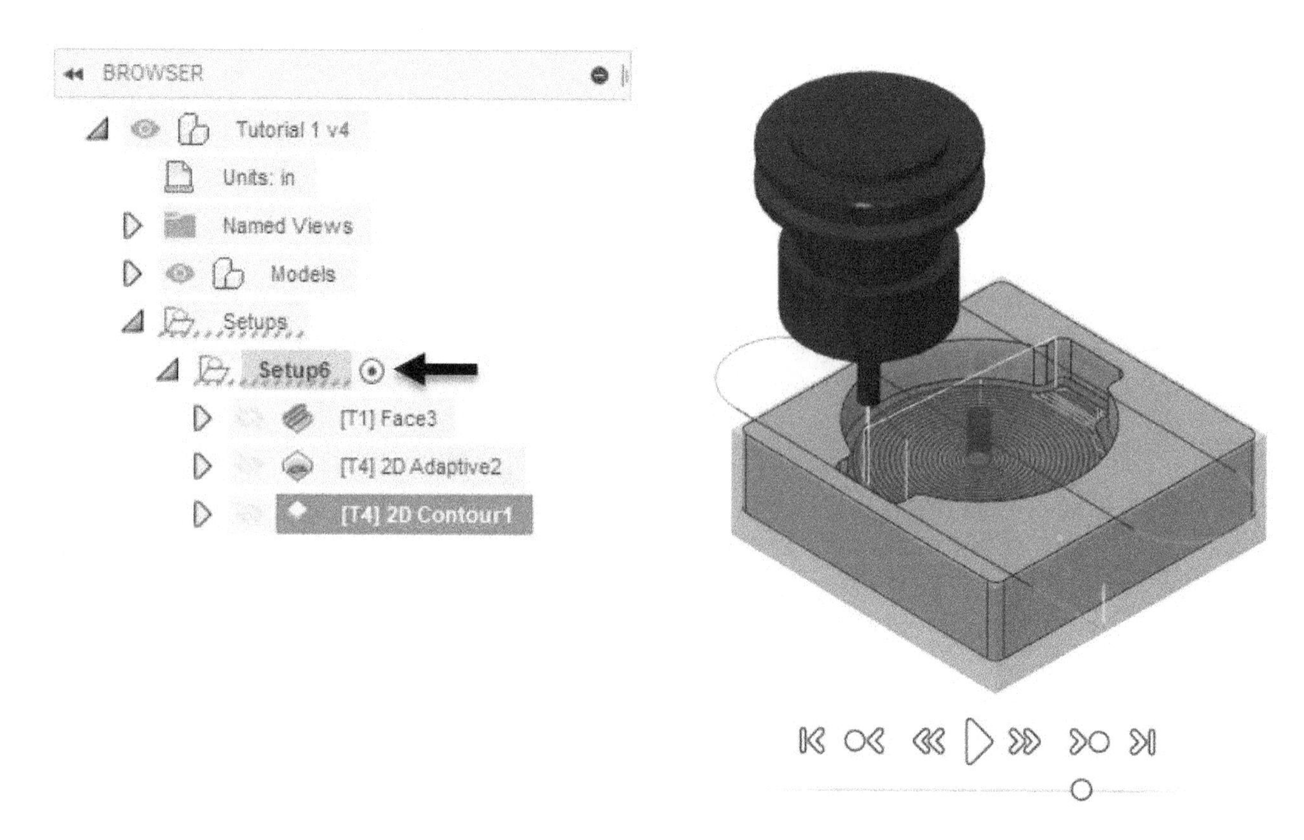

15. Click the **Close** button on the **Simulate** dialog.

Machining the outer faces of the model

1. On the Toolbar, click **Milling > 2D > 2D Adaptive Clearing**.
2. On the **2D Adaptive** dialog, click the **Select** button next to the **Tool** option.
3. On the **Select Tool** window, select the **Tutorial - Inch** folder from the **Fusion 360 Library**.
4. Select the **3-Ø1/2" L2" (flat end Mill)** tool from the tool list.
5. Click the **Select** button located at the bottom-right corner.
6. Select the bottom edge chain of the model, as shown.

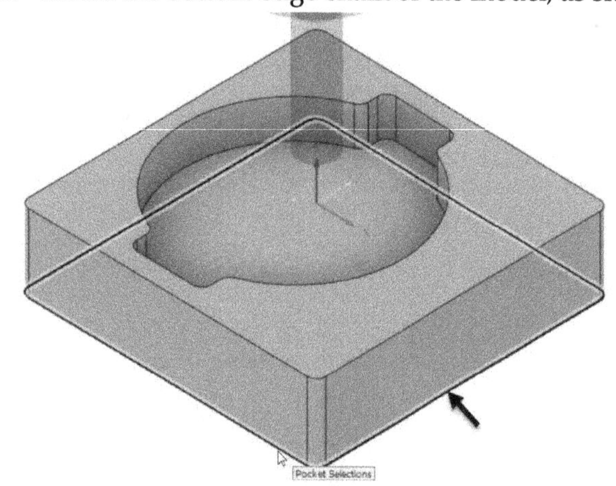

7. Click the **Heights** tab on the **2D Adaptive** dialog.
8. In the **Top Height** section, select **From > Model top**. The tool starts cutting from the top face of the model.

9. Leave the **Offset** value to **0**.
10. In the **Bottom Height** section, select **From > Selected contours(s)**. The tool removes the material up to the selected face.
11. Leave the **Offset** value to **0**.
12. Click the **Passes** tab on the **2D Adaptive** dialog.
13. Right-click in the **Optimal Load** box and select **Edit Expression**.
14. Change the expression to **tool_diameter*0.25**.
15. Click **OK** on the **Expression** dialog; the **Optimal Load** value is changed to 0.125.
16. Check the **Multiple Depths** option.
17. Change the **Maximum Roughing Stepdown** value to **0.4**.
18. Change the **Radial Stock to Leave** value to **0.01**.
19. Change the **Axial Stock to Leave** value to **0.00**.
20. Click **OK** on the **2D Adaptive** dialog to create the adaptive clearing milling operation.
21. In the Browser, click the right mouse button on the newly created **2D Adaptive** operation and select **Create Derived Operation > 2D Milling > 2D Contour**.
22. Click **OK**.

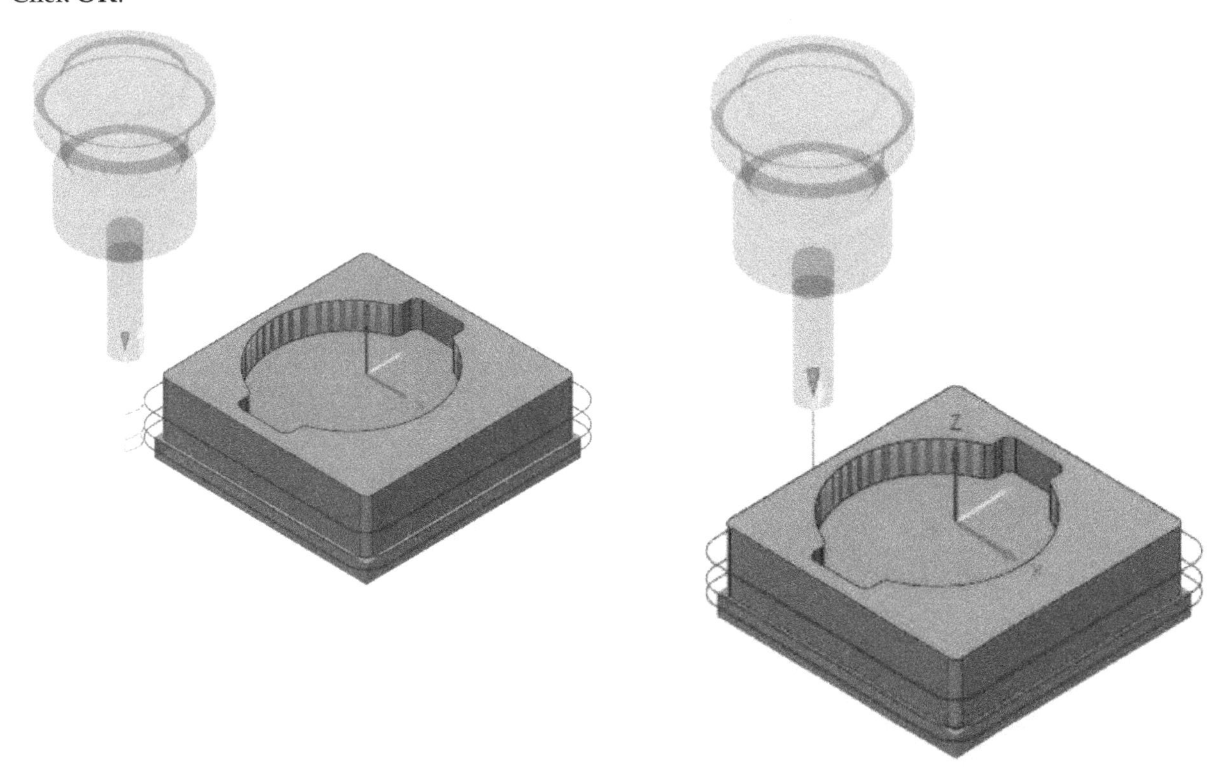

Post Processing

1. In the Browser, click the right mouse button on the **Setup** and select **Edit**.
2. Click the **Post Process** tab on the **Setup** dialog.

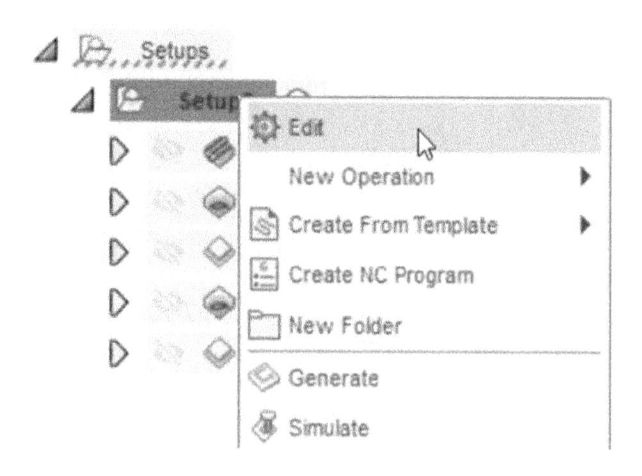

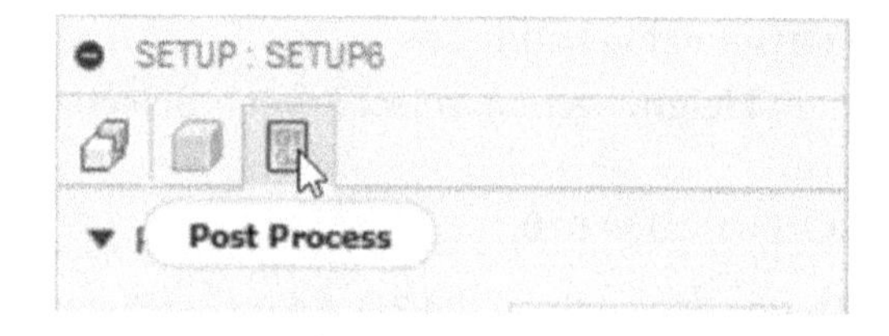

3. Type **1234** in the **Program Name/Number** box.
4. Type **JOB 1** in the **Program comment** box. Next, click **OK**.
5. Select the **Setup** from the Browser.
6. On the Toolbar, click **Milling > Actions > Post Process.**

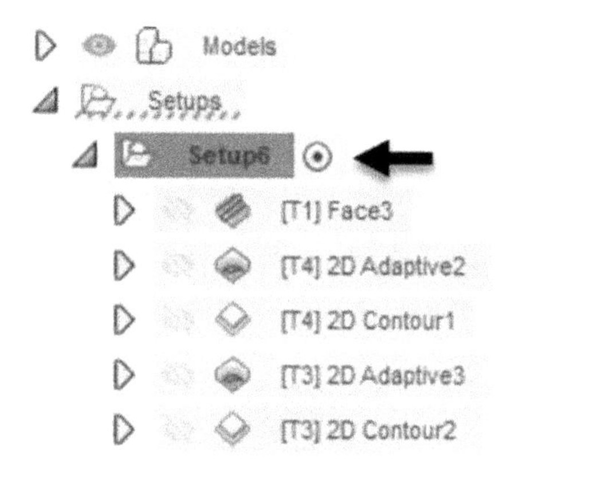

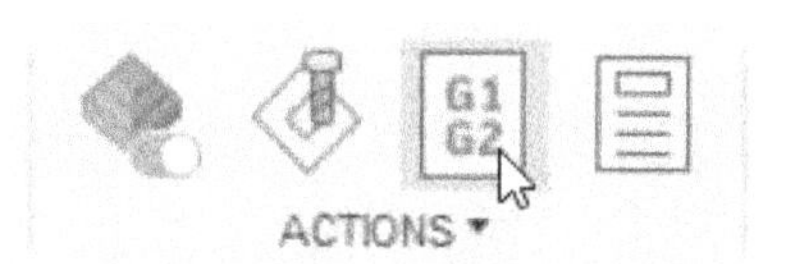

The **Post Process** dialog appears.

7. On the **Post Process** dialog, specify the configuration file's location using the **Browse** button next to the **Configuration folder** box. You can also leave it to the default location.

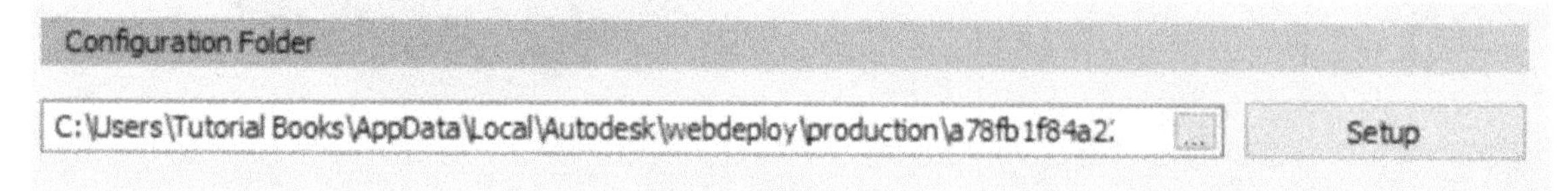

8. Select the vendor from the Vendor drop-down list. For example, select Autodesk.
9. Select **Milling** from the **Capabilities** drop-down.
10. Select **KOSY/kosy** from the **Configuration** drop-down.

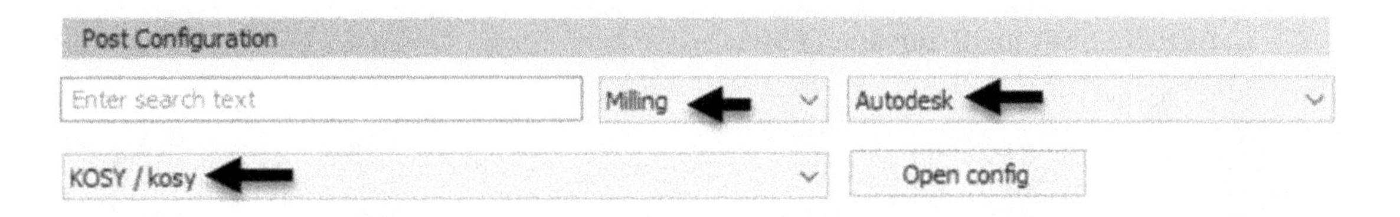

11. Specify the location of the output file using the button next to the **Output folder** box.

    Notice the properties in the Properties table located at the bottom right corner. The values change based on the configuration file.

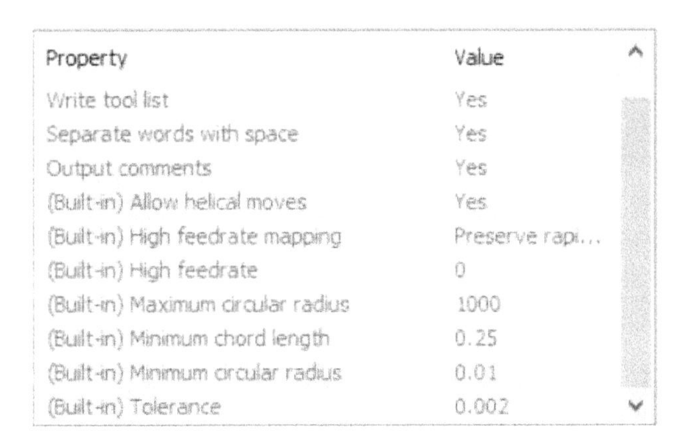

Property	Value
Write tool list	Yes
Separate words with space	Yes
Output comments	Yes
(Built-in) Allow helical moves	Yes
(Built-in) High feedrate mapping	Preserve rapi...
(Built-in) High feedrate	0
(Built-in) Maximum circular radius	1000
(Built-in) Minimum chord length	0.25
(Built-in) Minimum circular radius	0.01
(Built-in) Tolerance	0.002

12. Click the **Post** button.
13. Click the **Save** button. The G-code is generated and opened in the **Visual Studio Code** editor.

Creating Setup Sheets

1. In the Browser, select the **Setup**.
2. On the Toolbar, click **Milling > Actions > Setup Sheet**.

3. Specify the location of the Setup sheet and click the **Select Folder** button.

The setup sheet is opened in the internet browser. It displays the program name, job name, job description, job path, and a screenshot of the setup. The **Total** section displays information such as the total number of tools, operations, and maximum federate.

Setup Sheet for Program 1234

Program Comment: Job1
Job Description: Setup2
Document Path: Tutorial 1 (v2~recovered)

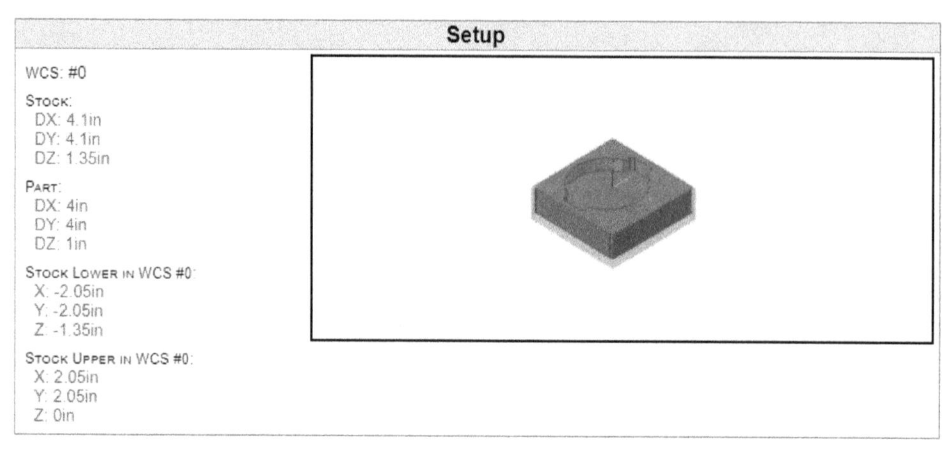

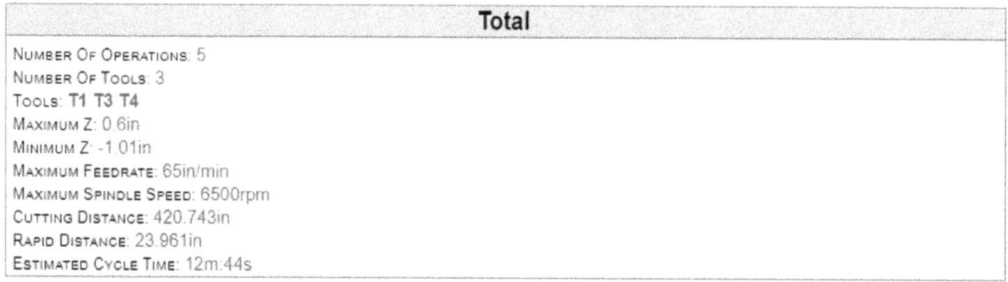

The **Tools** section displays the tools' specifications, such as tool type, diameter, and length.

The **Operations** section displays all the operations performed in the setup.

4. Save and close the design file.

Tutorial 2

In this tutorial, you perform the drilling operations.

Uploading the CAD Model

1. Download the **Tutorial 2** file.
2. On the Application Bar, click **File** drop-down > **Upload**.

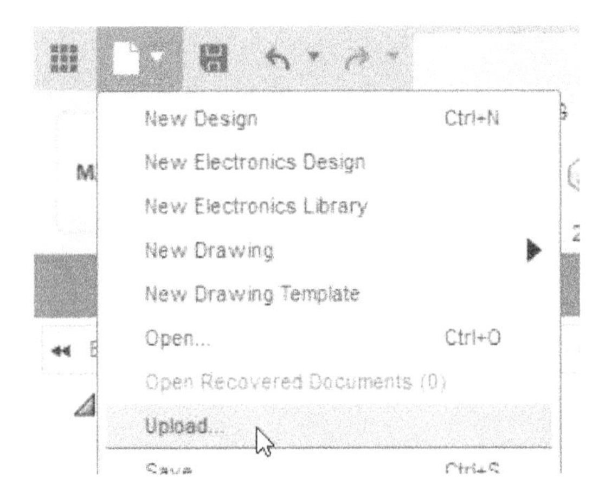

3. Click the **Change Location** option next to the **Location** box.
4. Select the **Fusion 360 CAM Overview** project from the Project list.
5. Click the **Select** button. Next, click the **Select Files** button.
6. Browse to the location of the downloaded file. Next, double-click on the Tutorial 2 file.
7. Click the **Upload** button. Next, close the **Job Status** dialog.
8. Click the **Show Data Panel** button located at the top left corner of the window.
9. Double-click on the **Fusion 360 CAM Overview** project to open it.
10. Double-click on the **Tutorial 2** file to open it. Next, close the **Data** panel.

Creating the Spot Drilling Operation

The spot drilling operation is performed before the actual drilling operation. The Spot drill tool is used to perform this operation.

1. On the Toolbar, click **Milling > Drilling > Drill**.
2. On the **Drill** dialog, click the **Select** button next to the **Tool** option.
3. On the **Select Tool** window, select the **Sample Tools – Inch** folder from the **Fusion 360 Library**.

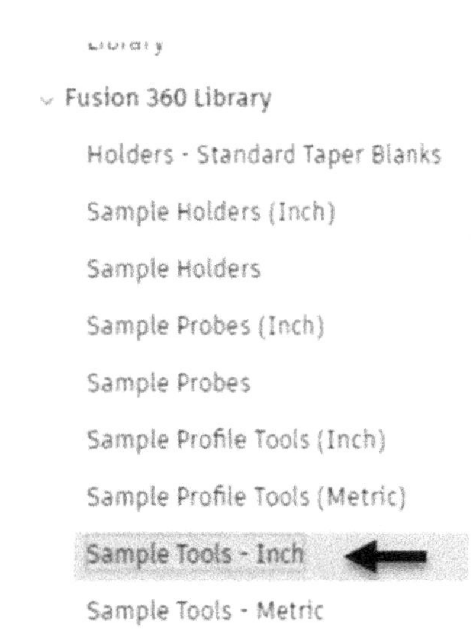

4. Select the **Ø5/8" L1.35" (5/8" Spot Drill)** tool from the tool list.
5. Select the **Aluminum – Drilling** option from the **Cutting Data** section.

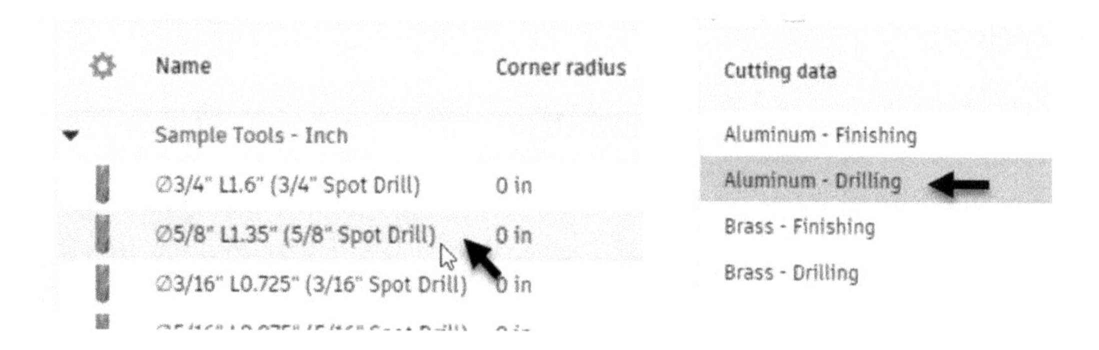

6. Click the **Info** tab on the top right corner. Notice that the **Tip angle** is set to 90 degrees.
7. Click the **Select** button located at the bottom-right corner. The Feed & Speed rates are automatically populated on the **Drill** dialog.

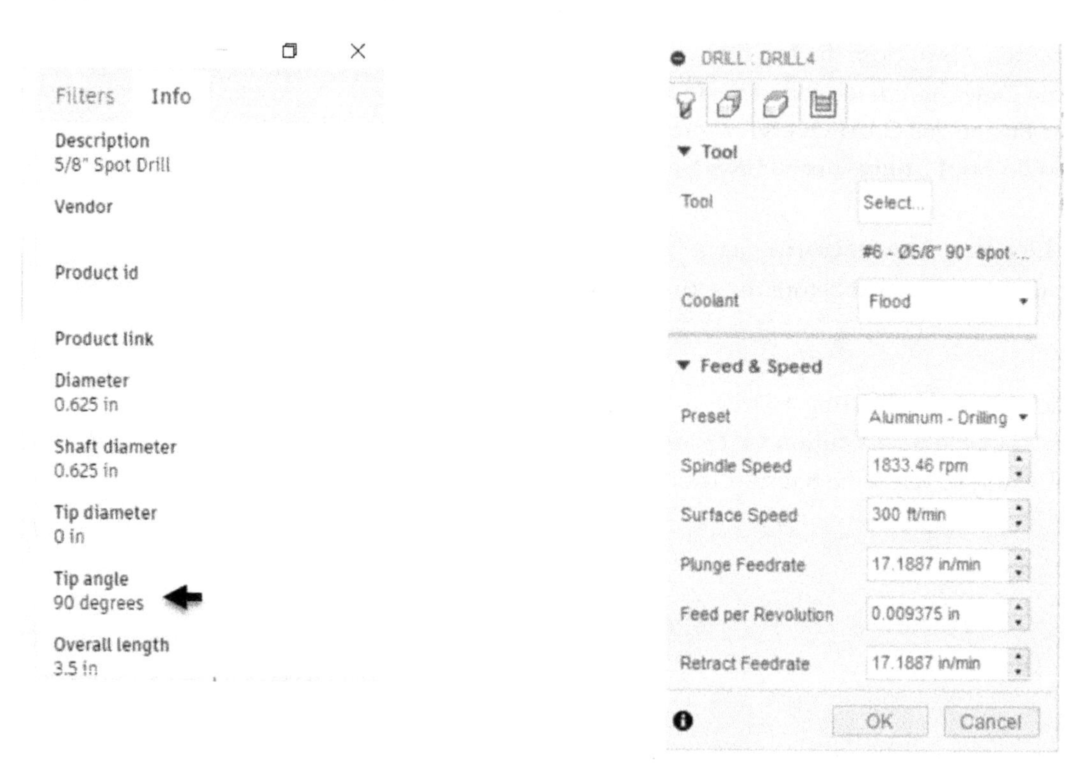

8. Click the **Geometry** tab on the **Drill** dialog.

Now, you need to select the holes to drill. There are three methods to drill the holes: **Selected faces, Selected points, Diameter range.**

The **Selected faces** option helps you select the holes by manually selecting the holes' cylindrical faces. If you want to select multiple holes of the same diameter, check the **Select Same diameter** option.

The **Selected points** option is used to specify the holes by selecting the circular edges of the holes.

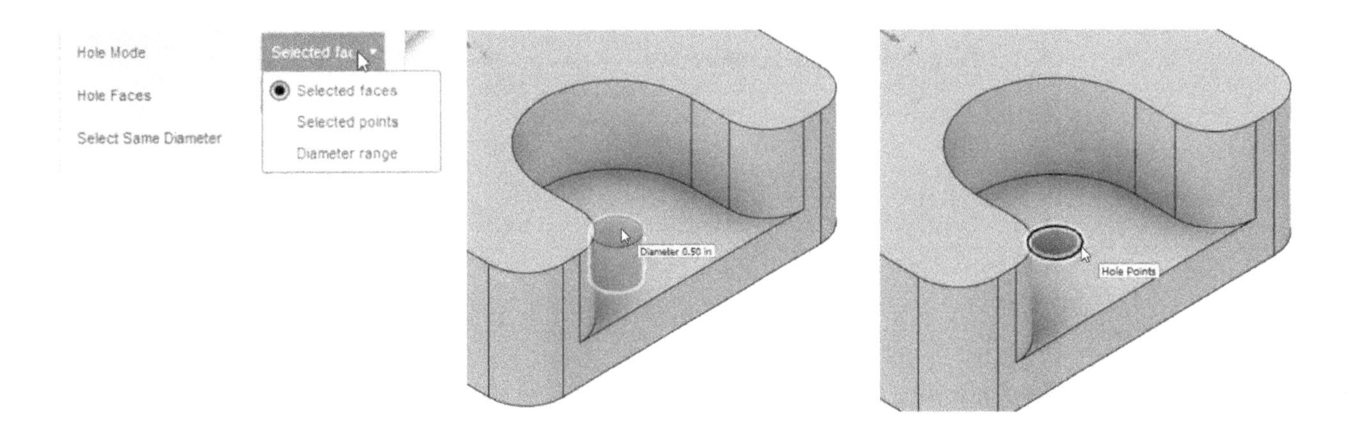

The **Diameter range** option is used to select the holes by specifying the diameter range.

9. Select the **Diameter range** option from the **Hole Mode** drop-down.
10. Type **0.4** and **0.5** in the **Minimum Diameter** and **Maximum Diameter** boxes, respectively. The holes in the specified range are selected automatically.

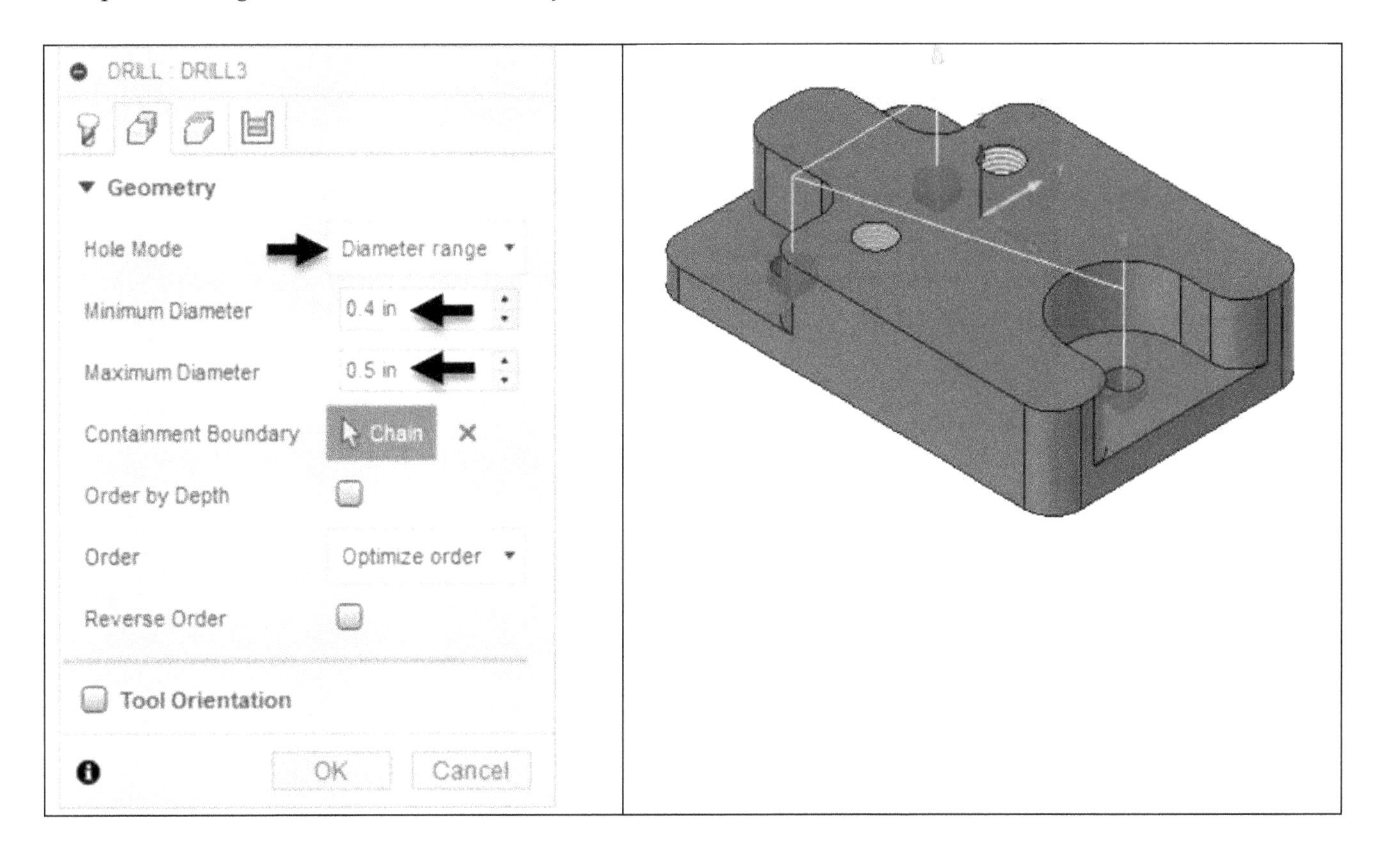

11. Click the **Heights** tab on the **Drill** dialog.

Next, you need to specify the center drilling depth. It is calculated based on the following formula.

$$Center\ Drilling\ Depth = \frac{Hole\ Diameter/2}{\tan(\frac{Tip\ angle}{2})}$$

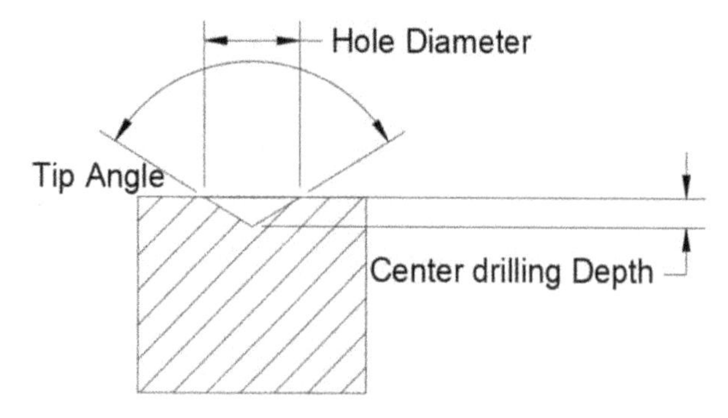

For this example, the Hole Diameter is **0.5**
Tip angle is **90**

$$Center\ Drilling\ Depth = \frac{0.5/2}{\tan(\frac{90}{2})} = 0.25$$

12. Under the **Bottom Height** section, select **From > Hole top.**
13. Type **-0.25** in the **Offset** box.
14. Click the **Cycle** tab on the **Drill** dialog. Next, select **Cycle Type > Drilling- rapid out**.
15. Click **OK** on the **Drill** dialog.
16. Select the **Drill [Rapid Out]** operation from the **Setup** section in the Browser. Next, simulate the selected operation using the **Simulate** tool.

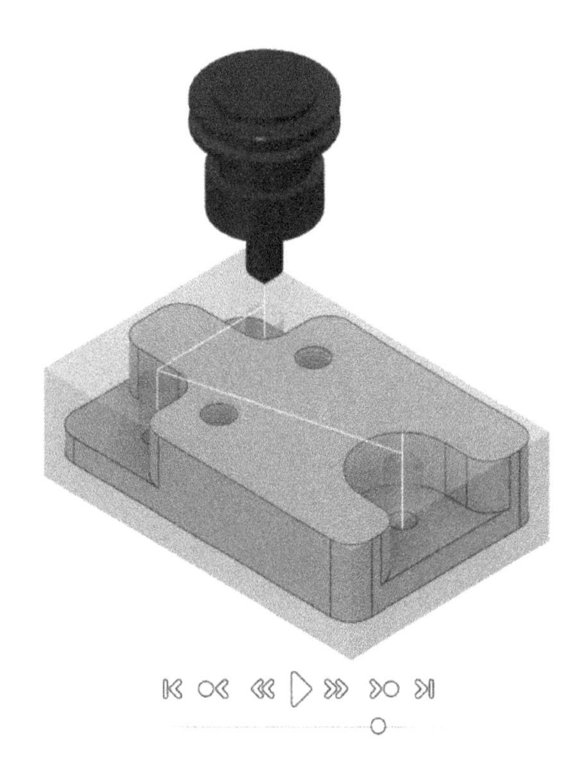

Creating the Through Drilling Operation

Now, you need to add another drilling operation to perform through drilling.

1. On the Toolbar, click **Milling > Drilling > Drill**.
2. On the **Drill** dialog, click the **Select** button next to the **Tool** option.
3. On the **Select Tool** window, select the **All** folder.
4. Click the **Filters** tab on the top-right corner. Next, select **Tool Category > Hole making**.

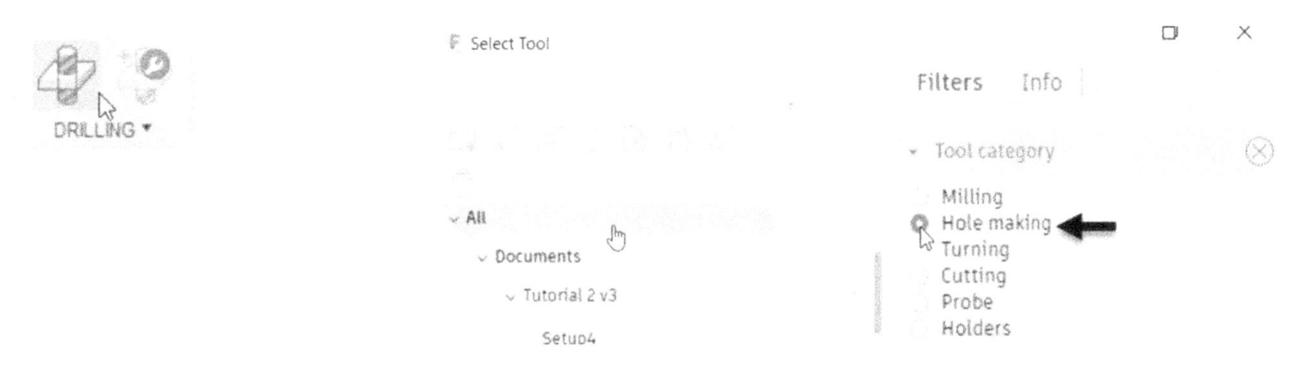

5. Select **Type > Drill**. Next, expand the **Diameter** section. Next, select **Equal** from the drop-down.
6. Type **0.5** in the box displayed next to the drop-down.
7. Select the **Ø1/2" L5.1" (1/2)** tool from the tool list.

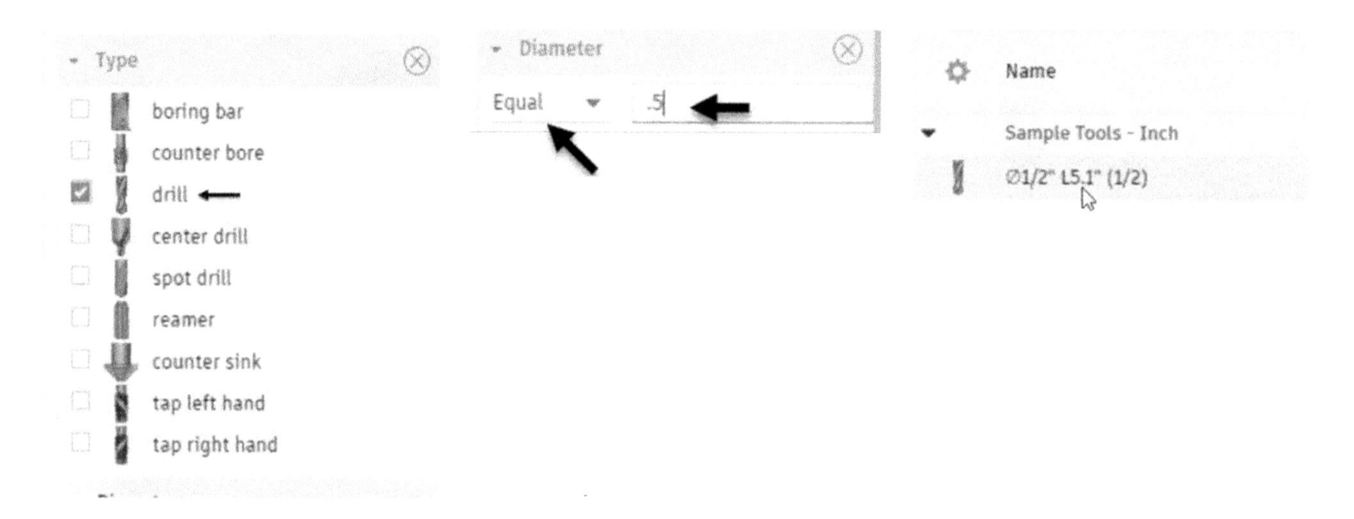

8. Select the **Aluminum - Drilling** option from the **Cutting Data** section.
9. Click the **Select** button located at the bottom-right corner. The Feed & Speed rates are automatically populated on the **Drill** dialog.
10. Click the **Geometry** tab on the **Drill** dialog.
11. Select the **Selected faces** option from the **Hole Mode** drop-down.
12. Select the cylindrical face of the hole, as shown.
13. Check the **Select Same Diameter** option; the holes with the same diameter of the selected hole are selected.

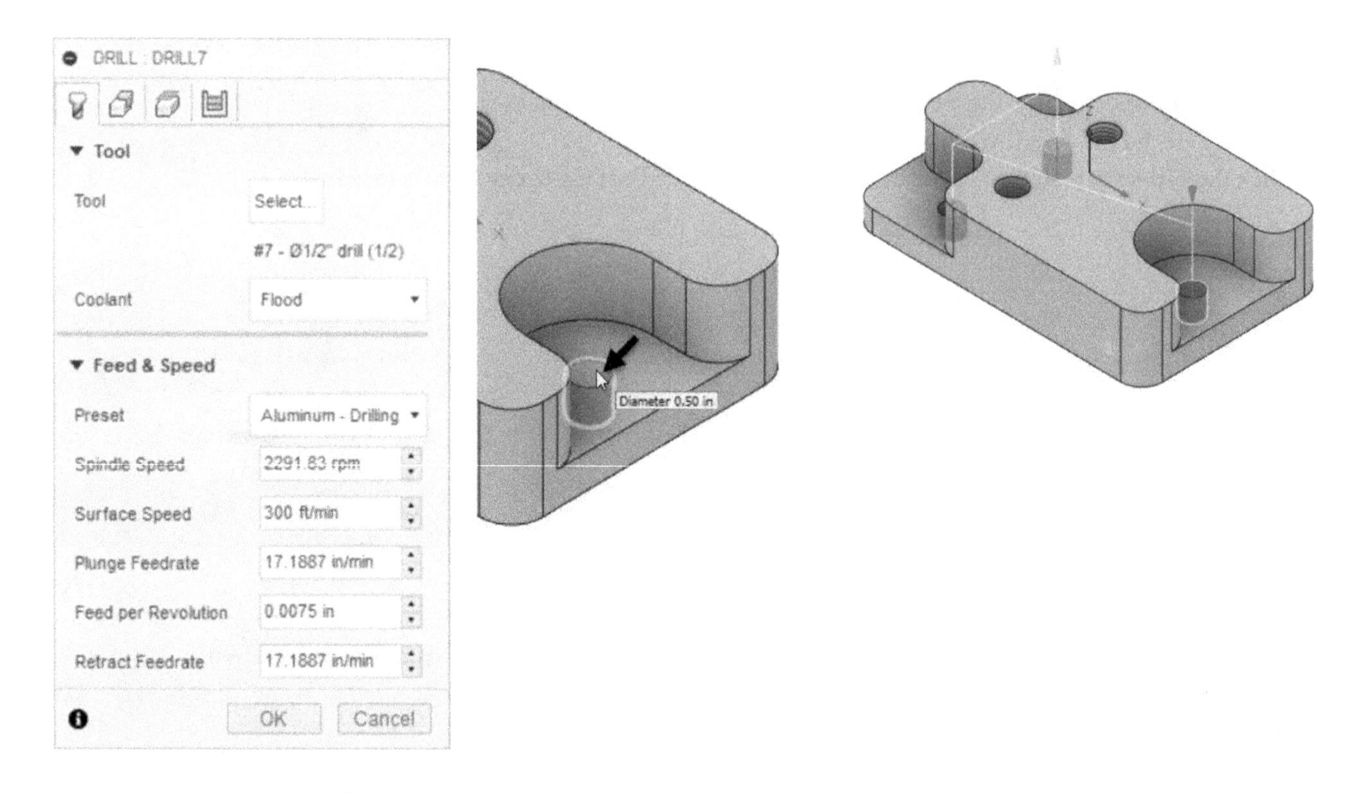

14. Click the **Heights** tab on the **Drill** dialog. Under the **Top Height** section, select **From > Hole top.**
15. Under the **Bottom Height** section, select **From > Hole Bottom.**

16. Check the **Drill Tip Through Bottom** option. You can change the view orientation and notice that the tooltip is outside the hole.
17. Click the **Cycle** tab on the **Drill** dialog. Next, select **Cycle Type > Deep drilling- full retract**.
18. Type 0.125 in the **Pecking Depth** box. Next, click **OK** on the **Drill** dialog.
19. Select the **Drill [Deep drilling]** operation from the Setup section in the Browser. Next, simulate the selected operation using the **Simulate** tool.

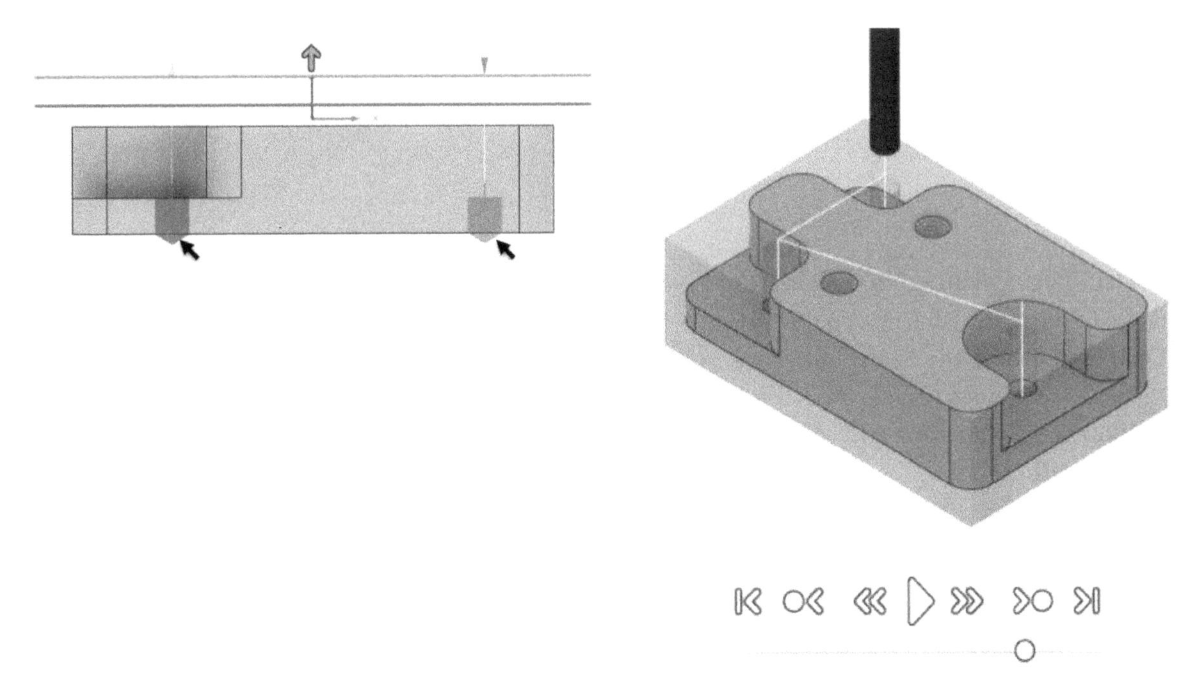

Creating the Derived Drilling Operations

1. Select the **Drill [Rapid Out]** operation from the **Setup** section in the Browser.
2. Right-click and select **Create Derived Operation > Drilling > Drill**.
3. Click the **Geometry** tab on the **Drill** dialog.
4. Select the **Diameter range** option from the **Hole Mode** drop-down.
5. Type **0.6** and **0.7** in the **Minimum Diameter** and **Maximum Diameter** boxes, respectively.
6. Click the **Heights** tab on the **Drill** dialog.
7. Under the **Bottom Height** section, select **From > Hole top.** Next, type **-0.303** in the **Offset** box.
8. Click the **Cycle** tab on the **Drill** dialog. Next, select **Cycle Type > Drilling- rapid out**.
9. Click **OK** on the **Drill** dialog.

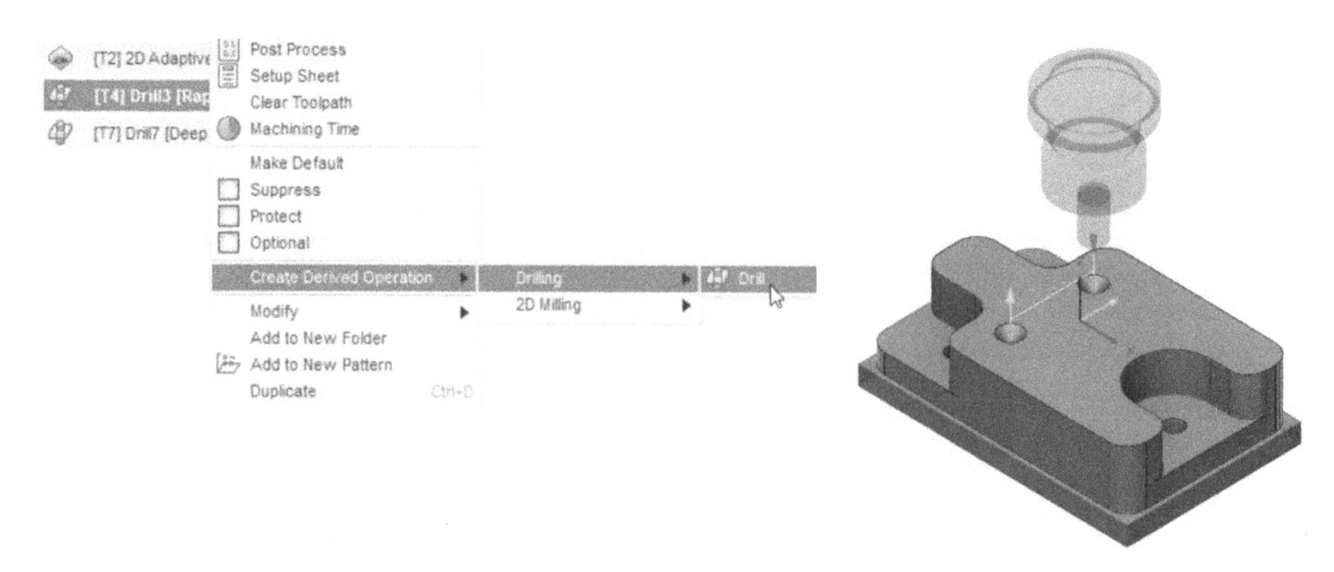

10. Select the **Drill [Deep drilling]** operation from the **Setup** section in the Browser.
11. Right-click and select **Create Derived Operation > Drilling > Drill**.
12. On the **Drill** dialog, click the **Select** button next to the **Tool** option.
13. On the **Select Tool** window, select the **All** folder.
14. Click the **Filters** tab on the top-right corner. Next., select **Tool Category > Hole making**.
15. Select **Type > Drill.**
16. Expand the **Diameter** section. Next, select **Greater than** from the drop-down.
17. Type **0.55** in the box displayed below the drop-down. Next, press ENTER.

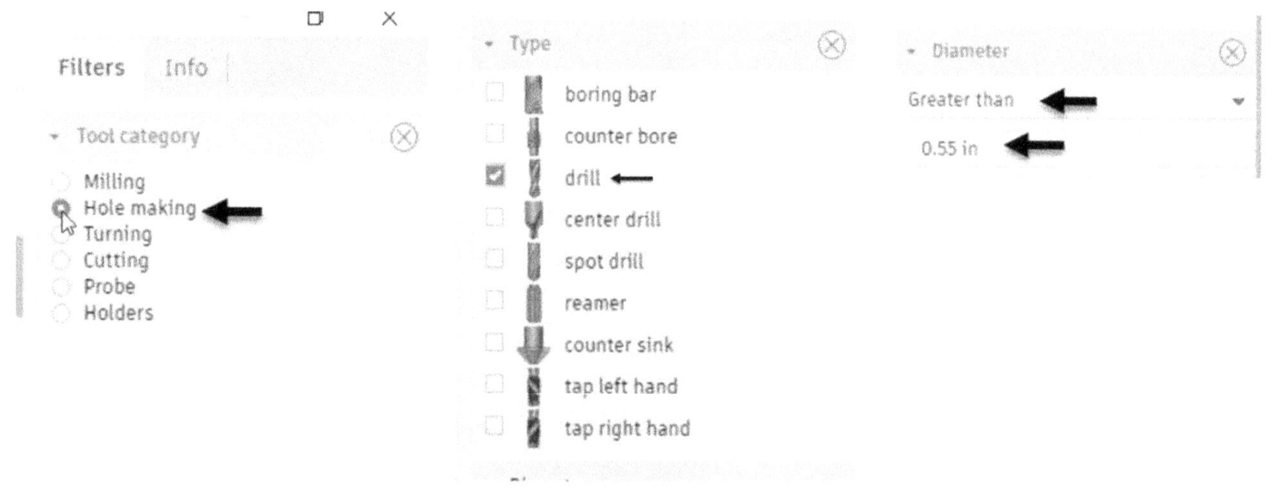

18. Select the **Ø19/32" L6.0375" (19/32)** tool from the tool list.
19. Select the **Aluminum - Drilling** option from the **Cutting Data** section.
20. Click the **Select** button located at the bottom-right corner. The Feed & Speed rates are automatically populated on the **Drill** dialog.
21. Click the **Geometry** tab on the **Drill** dialog.
22. Select the **Diameter range** option from the **Hole Mode** drop-down.
23. Type **0.6** and **0.7** in the **Minimum Diameter** and **Maximum Diameter** boxes, respectively.
24. Click the **Cycle** tab on the **Drill** dialog. Next, select **Cycle Type > Deep drilling- full retract**.
25. Type 0.15 in the **Pecking Depth** box. Click **OK** on the **Drill** dialog.

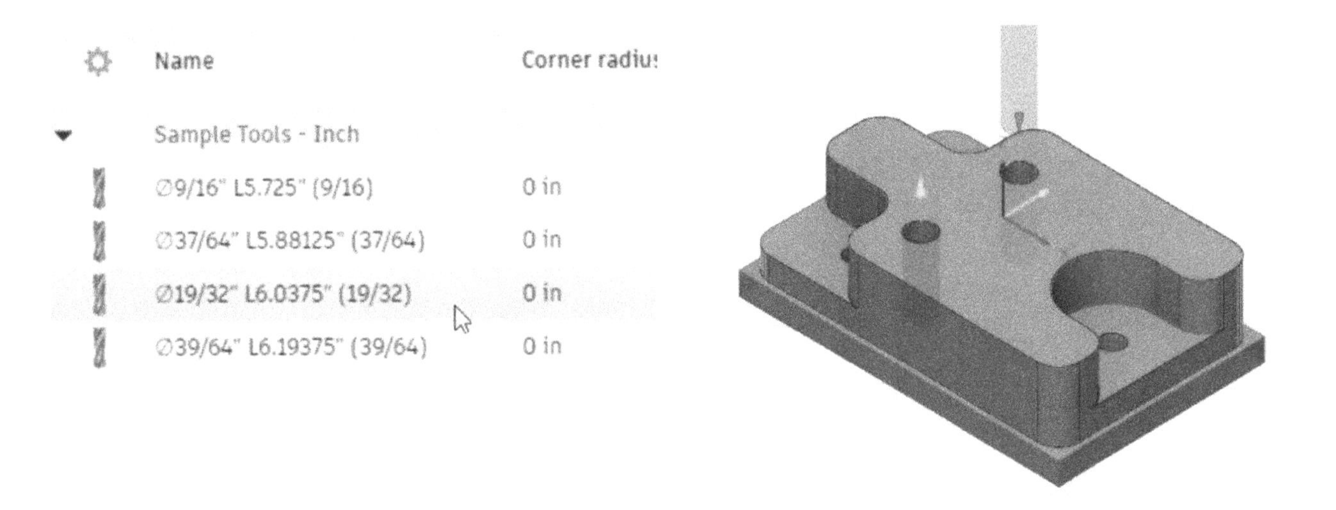

26. Select the **Drill [Deep drilling]** operation from the **Setup** section in the Browser. Next, simulate the selected operation using the **Simulate** tool.

Creating the Tapping Operations

1. Select the last **Drill [Deep drilling]** operation from the **Setup** section in the Browser.
2. Right-click and select **Create Derived Operation > Drilling > Drill**.
3. On the **Drill** dialog, click the **Select** button next to the **Tool** option.
4. On the **Select Tool** window, select the **All** folder.
5. Click the **Filters** tab on the top-right corner. Next, select **Tool Category > Hole making**.

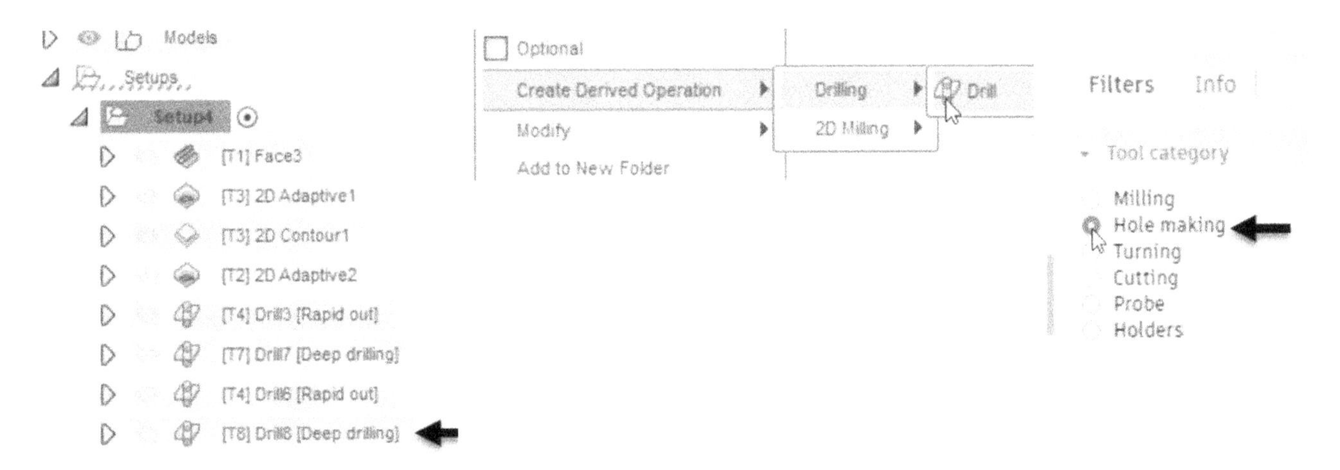

6. Select **Type > tap right hand**.
7. Expand the **Diameter** section. Next, select **Range** from the drop-down.
8. Type **0.6** and **0.7** in the minimum and maximum boxes, respectively. Next, press ENTER.
9. Select the **Ø11/16" L6.875" (11/16-24 UNEF)** tool from the tool list.

10. Click the **Select** button located at the bottom-right corner. The Feed & Speed rates are automatically populated on the **Drill** dialog.
11. Click the **Cycle** tab on the **Drill** dialog.
12. Select **Cycle Type > Tapping**. Next, click **OK** on the **Drill** dialog.

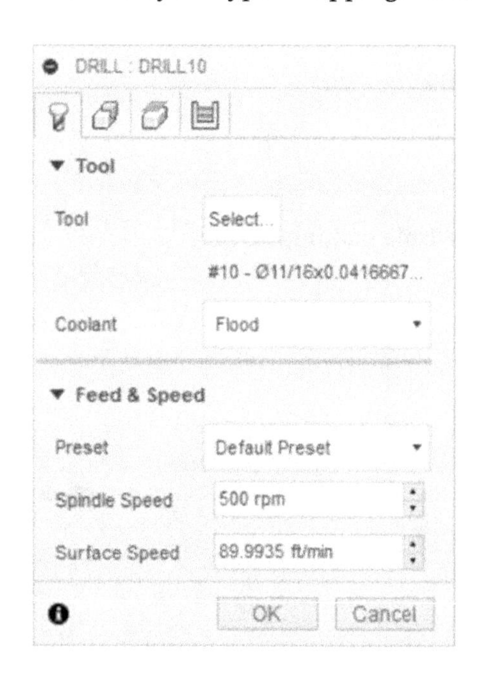

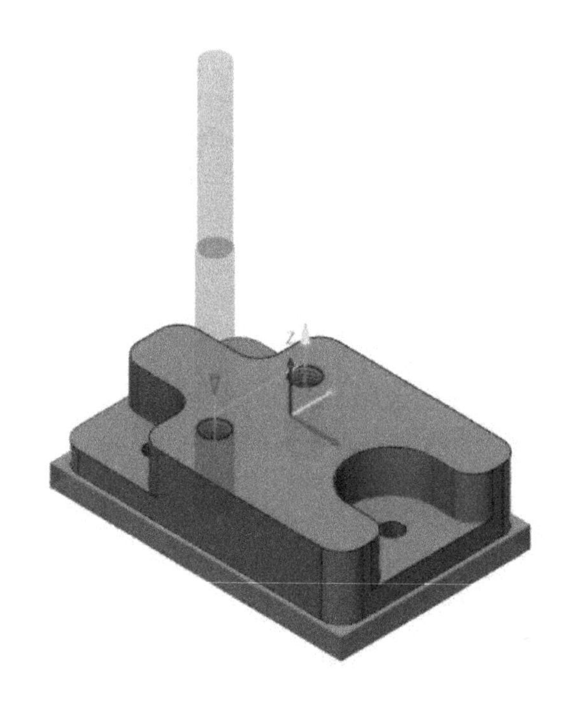

13. Press and hold the CTRL key and select all the drilling operations from the **Browser Setup** section. Next, simulate the selected operations using the **Simulate** tool.
14. Save and close the design file.

Tutorial 3

Uploading the CAD Model

1. Download the **Tutorial 3** file.
2. On the Application Bar, click **File** drop-down > **Upload**.

3. Click the **Change Location** option next to the **Location** box.
4. Select the **Fusion 360 CAM Overview** project from the **Project** list.
5. Click the **Select** button. Next, click the **Select Files** button.
6. Browse to the location of the downloaded file. Next, double-click on the Tutorial 3 file.
7. Click the **Upload** button. Next, close the **Job Status** dialog.
8. Click the **Show Data Panel** button located at the top left corner of the window.
9. Double-click on the **Fusion 360 CAM Overview** project to open it.
10. Double-click on the **Tutorial 3** file to open it. Next, close the **Data** panel.

Creating the Circular Milling Operation

The circular milling operation allows you to machine multiple bosses, grooves, and holes in a single operation.

1. On the Toolbar, click **Milling > 2D > Circular**.
2. On the **Circular** dialog, click the **Select** button next to the **Tool** option.
3. On the **Select Tool** window, select the **Sample Tools - Inch** folder from the **Fusion 360 Library**.
4. Select the **Ø5/8" L1.35" (5/8" Flat Endmill)** tool from the tool list.
5. Select **Aluminum - Finishing** from the **Cutting Data** section. Next, click the **Select** button.
6. Click the **Geometry** tab on the **Circular** dialog.
7. Select the circular faces of the outside and inside bosses, as shown.

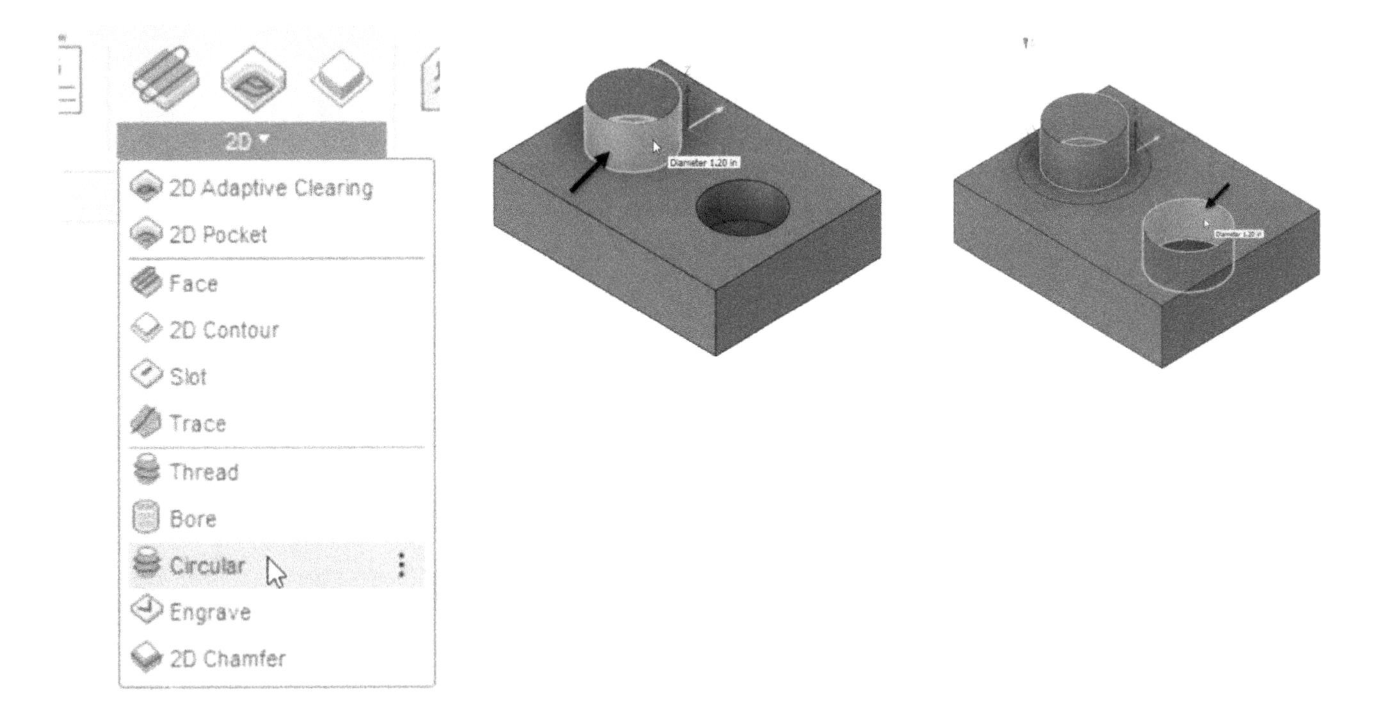

8. Click the **Passes** tab on the **Circular** dialog.
9. Click the right mouse button in the **Maximum Roughing Stepdown** box, and then select **Reset to Default**.
10. Select the **Compensation Type > Wear**.
11. Click the **Height** tab and type 0.04 in the **Offset** box available in the **Bottom Height** section.

12. Click the **Linking** tab on the **Circular** dialog.
13. Check the **Lead To Center** option; the lead-in/out will start from the center of the hole or pocket.

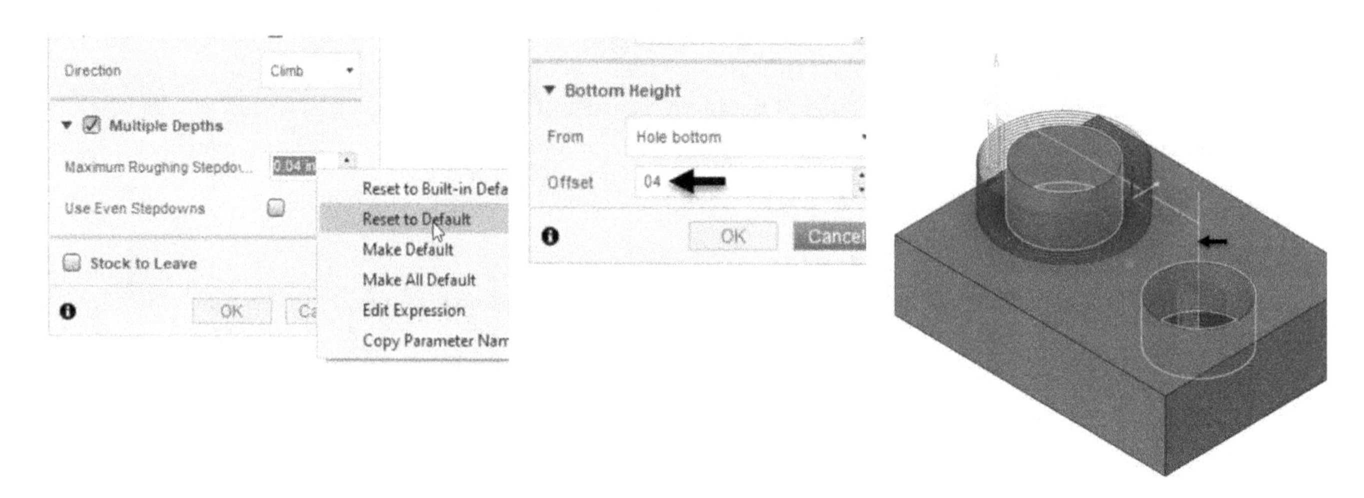

14. Click **OK** on the **Circular** dialog.

Creating the Thread Milling Operation

This operation generates a helical tool path for the machining of internal and external threads with thread mills.

1. On the Toolbar, click **Milling > 2D > Thread**.
2. On the **Thread** dialog, click the **Select** button next to the **Tool** option.
3. On the **Select Tool** window, select the **All** > **Local > Library** folder.

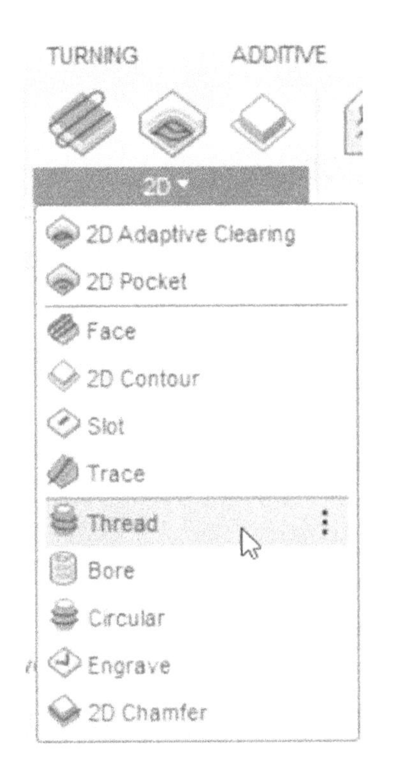

4. Click the **New Tool** icon. On the **New Tool** window, select **Milling > thread mill**.
5. Click the **Cutter** tab. Next, enter the following values in the **Geometry** section.

***Diameter:* 0.5**
***Shaft diameter:* 0.5**
***Overall length:* 3.5**
***Length below holder:* 2.5**
***Shoulder length:* 1.5**
***Flute length:* 1**
***Thread pitch:*0.1**
***Number of teeth:* 1**
***Thread profile angle:* 60 degrees**

6. Click the **Accept** button. Next, click the **Select** button.
7. Click the **Geometry** tab on the **Thread** dialog. Next, select the circular face of the boss, as shown.

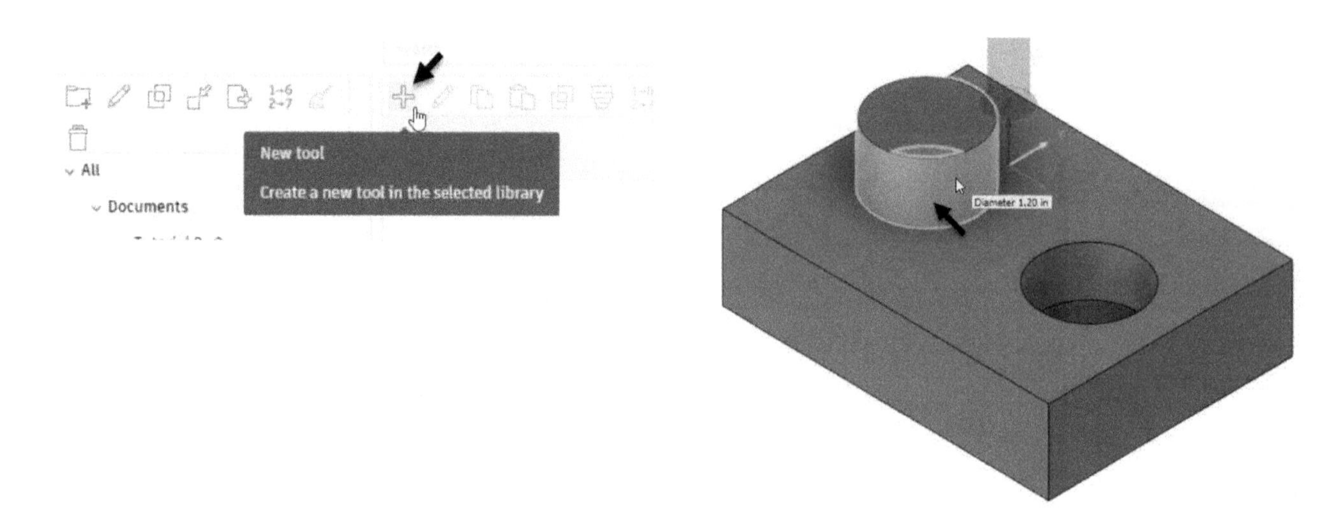

8. Click the **Heights** tab. Next, type **0.1** in the **Offset** box available in the **Bottom Height** box.
9. Click the **Passes** tab on the **Thread** dialog. Next, type **0.08** in the **Pitch Diameter Offset** box.
10. Select **Compensation Type > Wear**. Next, click **OK** to create the thread milling operation.

Tutorial 4

Creating the Slot Operation

This operation generates a tool path along the centerline of a slot.

1. Download the **Tutorial 4** file and upload it to Fusion 360.
2. Open the uploaded file.
3. On the Toolbar, click **Change Workspace** drop-down > **Manufacture**.
4. On the Toolbar, click **Milling** > **2D** > **Slot**.
5. On the **Slot** dialog, click the **Select** button next to the **Tool** option.
6. On the **Select Tool** window, select the **Sample Tools – Inch** folder from the **Fusion 360 Library**.
7. Select the **Ø1/4" L0.85" (1/4" Flat Endmill)** tool from the tool list.

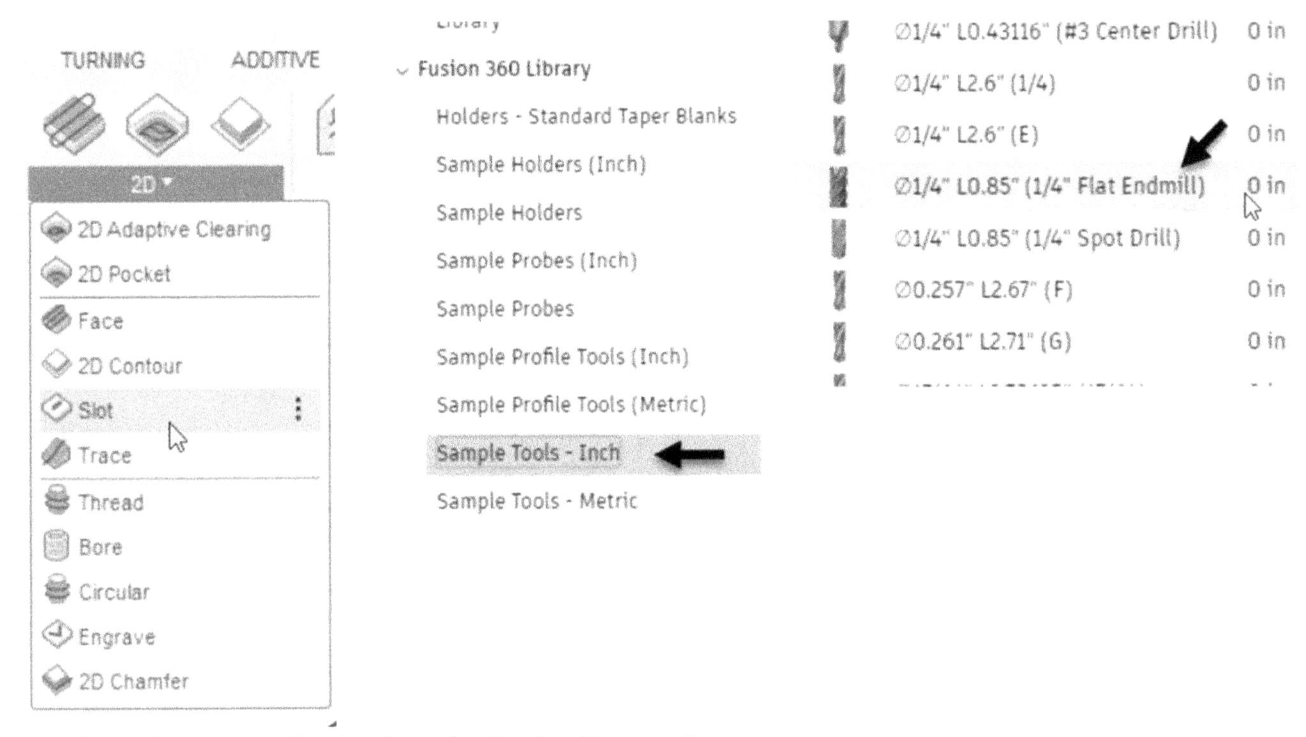

8. Select **Aluminum – Slotting** from the **Cutting Data** section.
9. Click the **Select** button located at the bottom-right corner.
10. Click the **Geometry** tab on the **Thread** dialog. Next, select the bottom face of the slot, as shown.

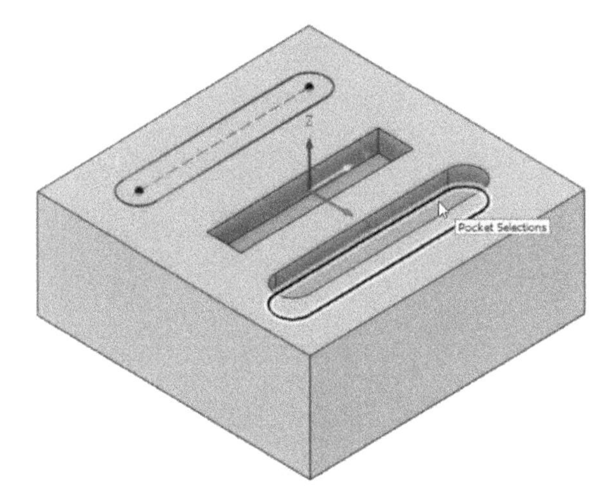

The selected slot is 0.5 inches wide. At the same time, the selected tool is of a 0.25-inch diameter. The 0.25-inch tool will not create a pocket on the selected contour. It will just remove the material by moving back and forth along the length of the selected slot. As a result, a 0.25-inch slot will be created.

11. Click the **Heights** tab. In the **Top Height** section, select **From > Model top**.
12. Click the **Passes** tab on the **Slot** dialog. On the **Passes** tab, you can specify the **Tolerance**, **Backoff Distance**, **Tangent Fragment Extension Distance**.

Backoff Distance

The distance to move away from the last cut, before retracting the tool.

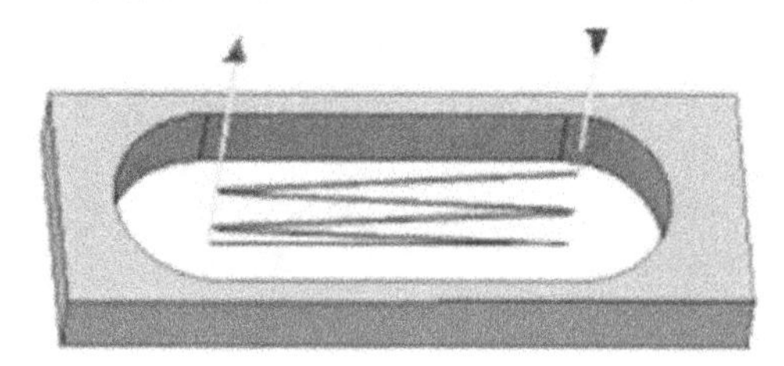

Distance 0.0 in

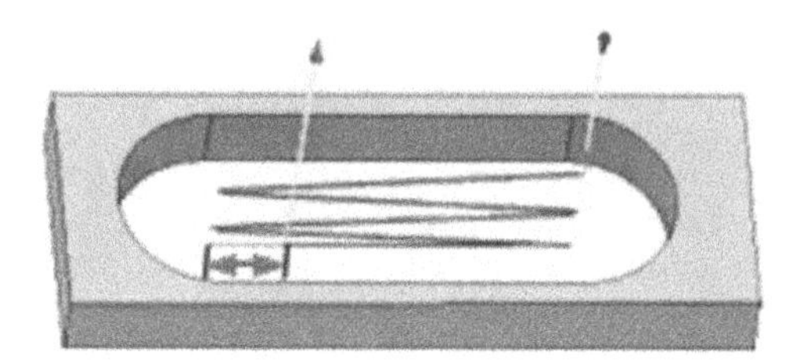

Distance .200in

Tangential Fragment Extension Distance

Used on open contours to extend the beginning and end of the calculated toolpath. This creates a tangent linear extension based on the angle of the start and endpoints. This extension can be used in combination with the **Geometry - Tangential Extension Distance**.

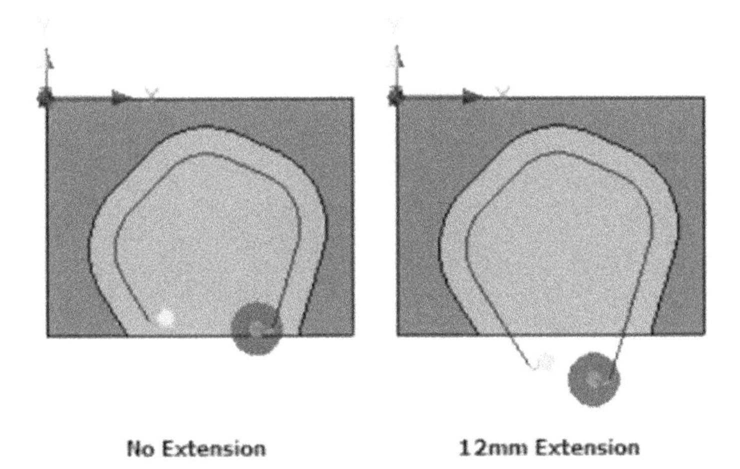

No Extension **12mm Extension**

Overlaping extensions of a single chain, will not be trimmed.

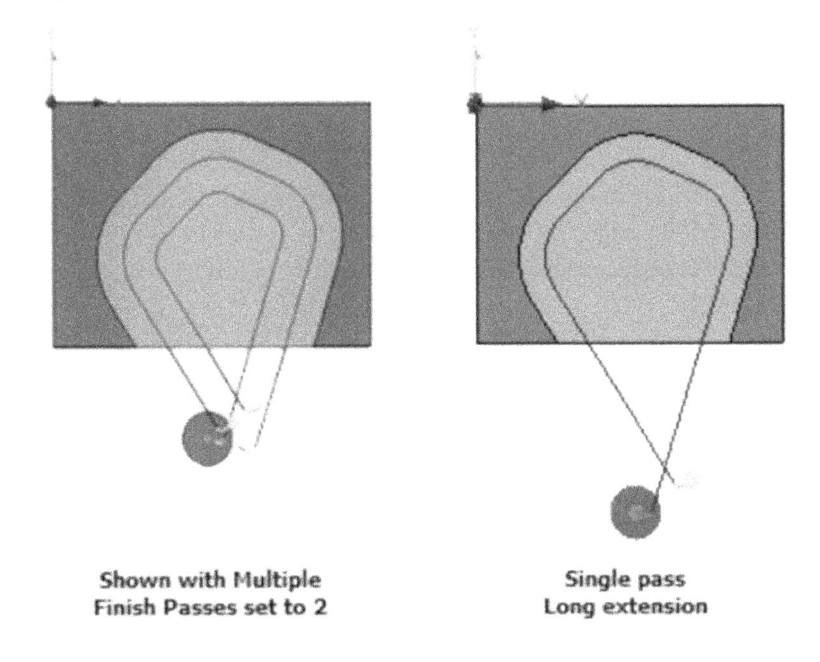

Shown with Multiple Finish Passes set to 2 **Single pass Long extension**

Note: You can use the Stock Contours option to force the toolpath past the defined Stock or a selected boundary. Great for irregular shapes. If you need a different extension for each end of the cut, you can use **Geometry Tab - Tangential Extension Distance.**

13. Click **OK** to create the slot operation.

Notice that the existing slot defines the length of the slot operation. At the same time, the tool diameter defines the slot width.

Likewise, create a slot operation on the slot with flat ends. Notice that the slot is created with curved ends.

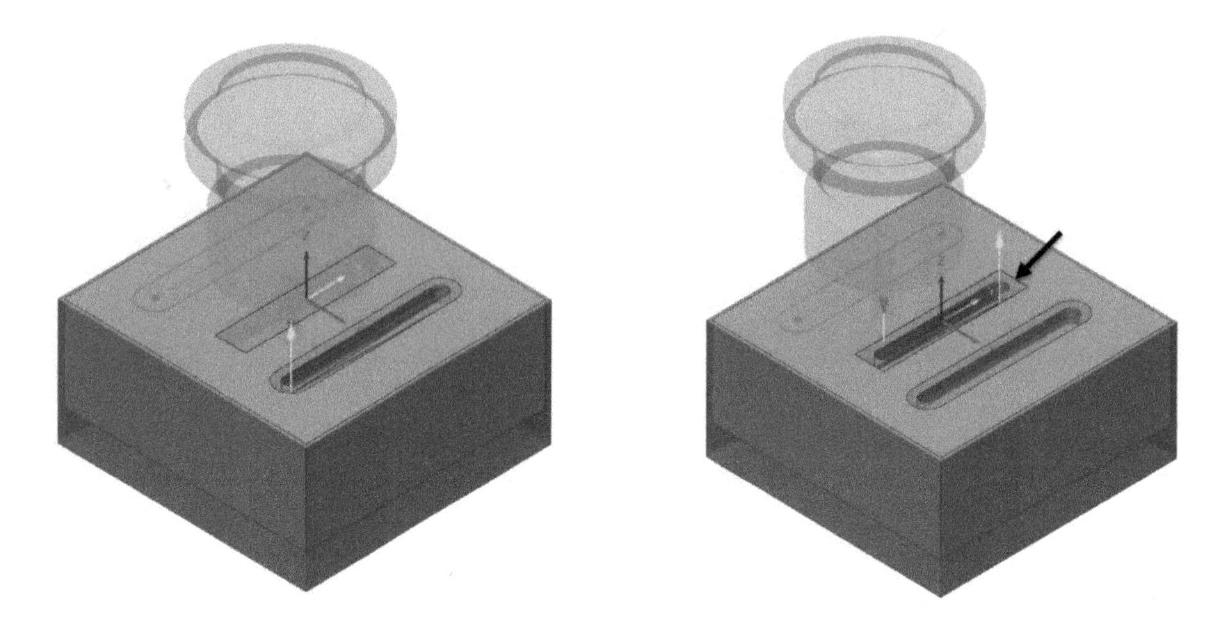

14. On the Toolbar, click **Milling** > **2D** > **Slot**.
15. Click the **Geometry** tab on the **Thread** dialog. Next, select the sketched slot, as shown.
16. Click the **Heights** tab.
17. In the **Top Height** section, select **From > Model top**. The tool starts cutting from the top face of the model.
18. Leave the **Offset** value to **0**. In the **Bottom Height** section, select **From > Selected contours(s)**.
19. Set the **Offset** value to **-0.25**. Next, click **OK**.
20. Hide the **Model** node in the Browser.
21. Simulate the newly created slot operation and notice that the slot is created up to a depth of 0.25 inches.

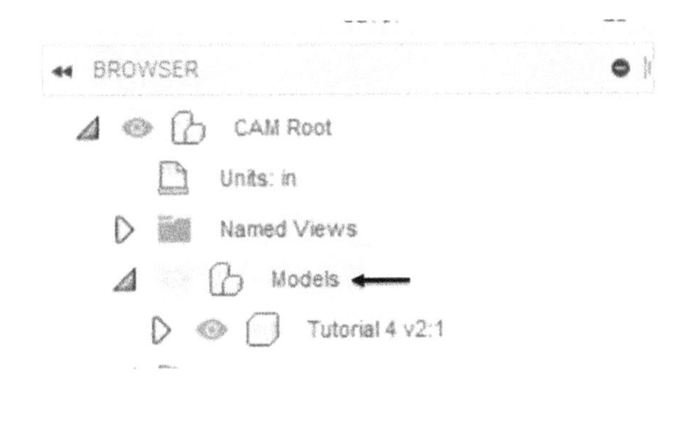

Tutorial 5

Upload the Design file and create the Facing and 2D Adaptive Clearing Operations

1. Download the **Tutorial 5** file, and then upload it to Fusion 360. Next, open the uploaded file.
2. On the Toolbar, click **Change Workspace** drop-down > **Manufacture**.
3. Create the Face and 2D Adaptive Clearing Operations.

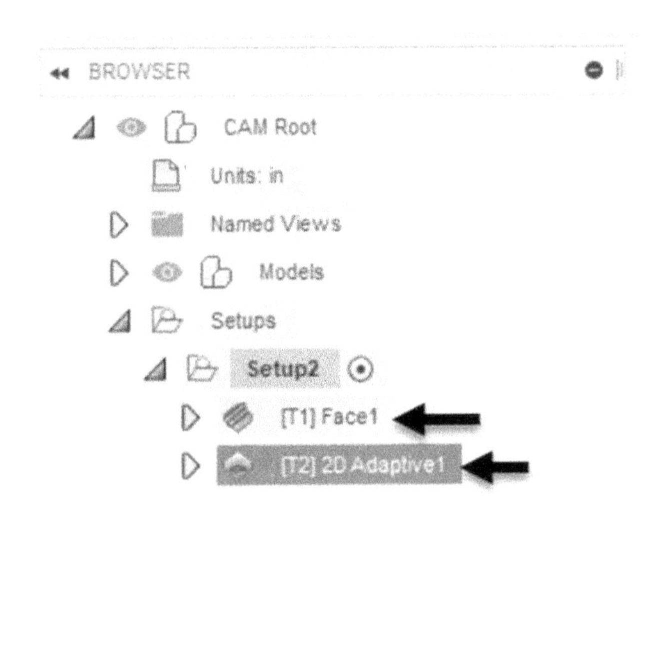

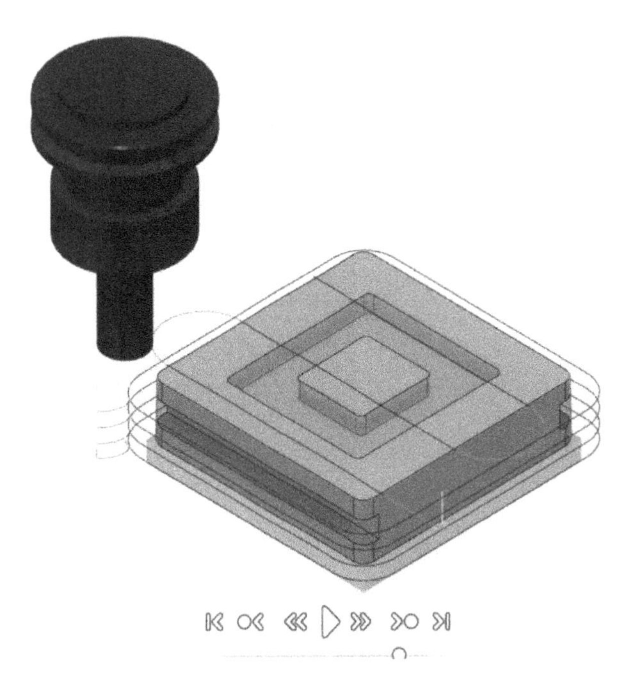

Creating the 2D Pocket Operation

In this section, you will create a 2D Pocket operation.

1. On the Toolbar, click **Milling > 2D > 2D Pocket**.
2. On the **2D Pocket** dialog, click the **Select** button next to the **Tool** option.
3. On the **Select Tool** window, select the **Tutorial - Inch** folder from the **Fusion 360 Library**.
4. Select the **4-Ø1/4" L1" (flat end Mill)** tool from the tool list.

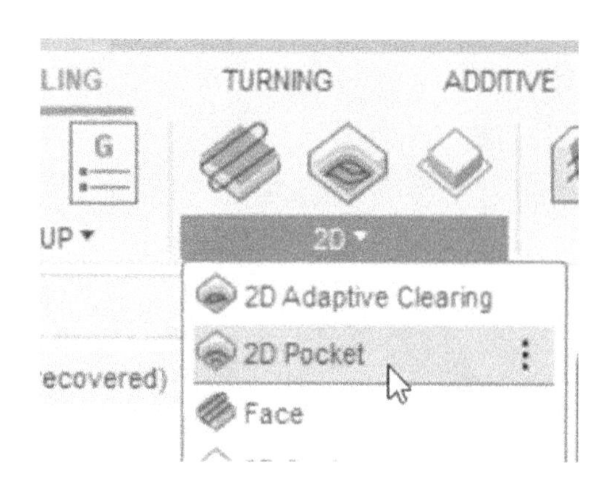

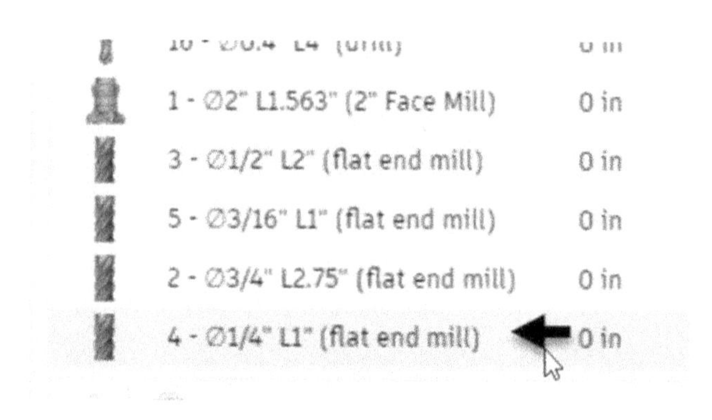

5. Click the **Select** button located at the bottom-right corner.
6. Click the **Geometry** tab on the **2D Pocket** dialog.
7. Click on the floor face of the pocket, as shown. The selected face is turned blue, which indicates the region to be cut.
8. Click the **Heights** tab on the **2D Pocket** dialog.
9. In the **Top Height** section, select **From > Model top**. The tool starts cutting from the top face of the model.
10. Leave the **Offset** value to **0**.
11. In the **Bottom Height** section, select **From > Selected contours(s)**. The tool removes the material up to the selected face.
12. Leave the **Offset** value to **0**.
13. Click the **Passes** tab on the **2D Pocket** dialog. Next, check the **Finishing Passes** option.

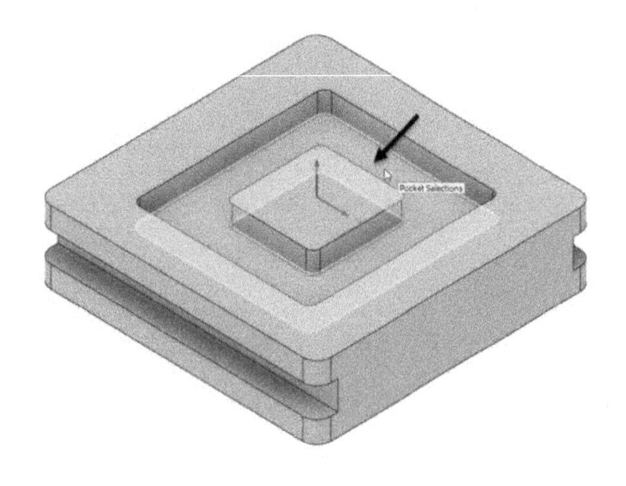

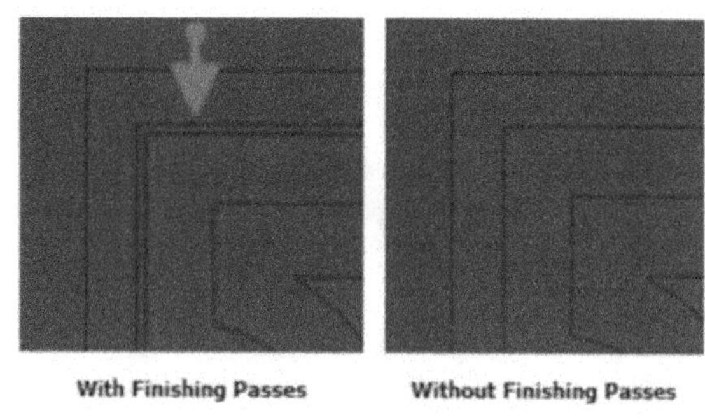

14. Type 0.2 in the **Maximum Stepover** box.

Maximum Stepover

Specifies the maximum horizontal stepover between passes.

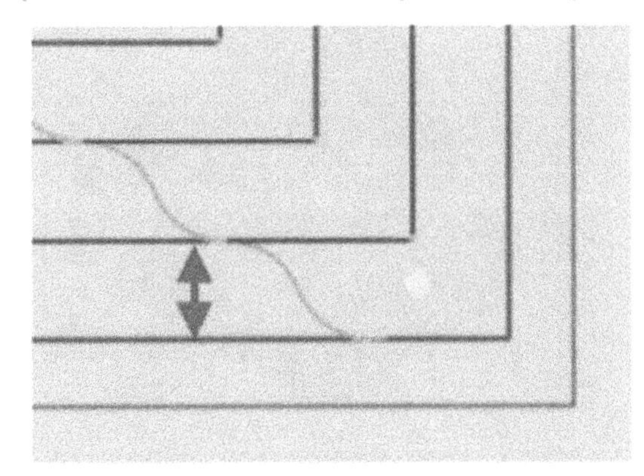

NOTE: This is NOT the same as optimal load settings with adaptive clearing paths. With legacy 2D roughing, the tool still sees full cutter engagement when transitioning from one pass to the next.

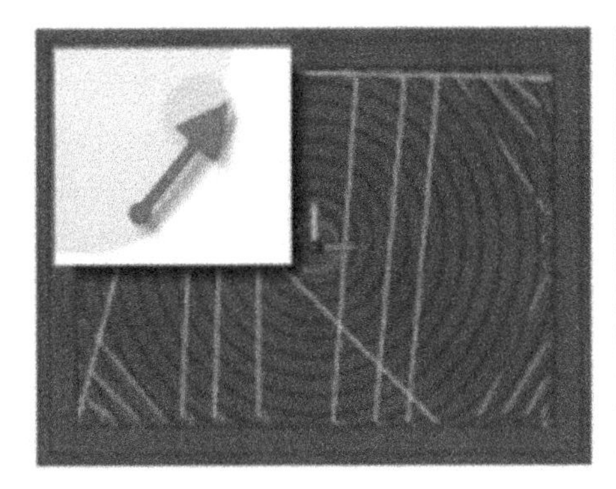

Adaptive Clearing **Legacy 2D Clearing**

15. Check the **Multiple Depths** option. Next, change the **Maximum Roughing Stepdown** value to **0.25**.
16. Check the **Finish Only at Final Depth** option. Next, check the **Use Even Stepdowns** option.
17. Change the **Radial Stock to Leave** value to **0.01**. Next, change the **Axial Stock to Leave** value to **0.00**.
18. Click the **Linking** tab on the **2D Pocket** dialog.
19. Type 0.1 in the **Safe Distance** box. The **Safe Distance** should be less than the **Feed Height**.
20. Click **OK** on the **2D Pocket** dialog to create the 2D pocket operation.

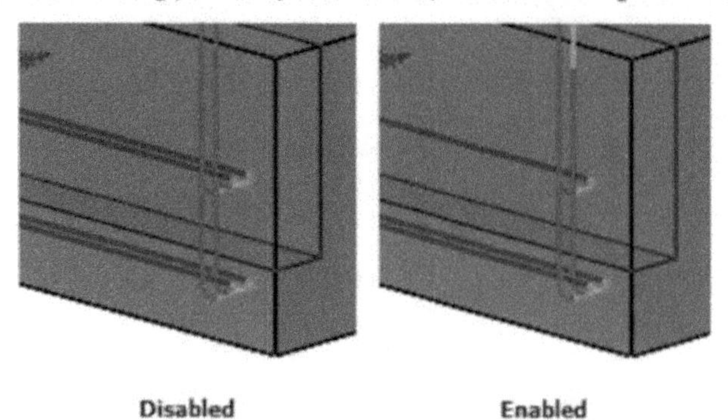

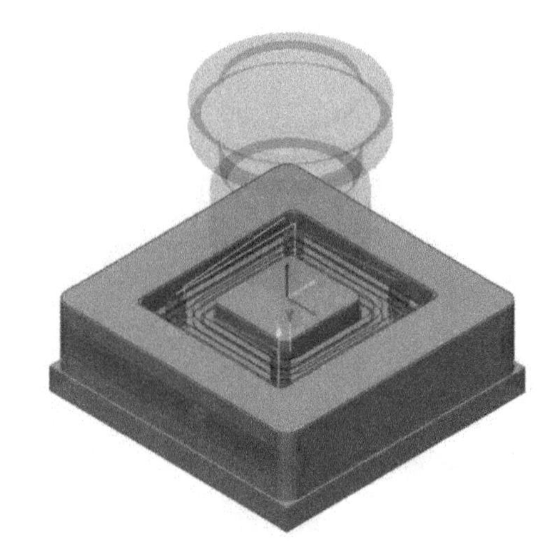

Milling the Slots on the Side Faces

1. On the Toolbar, click **Milling > Setup > Setup.** On the **Setup** dialog, select **Operation Type > Milling**.
2. Click the **Stock** tab on the **Setup** dialog. Next, select **Mode > Relative size box**.
3. Select **Stock Offset Mode > No additional stock**.
4. Click the **Setup** tab. Next, select **Orientation > Select Z Axis/plane & X axis**.
5. Select the side face to define the Z axis.

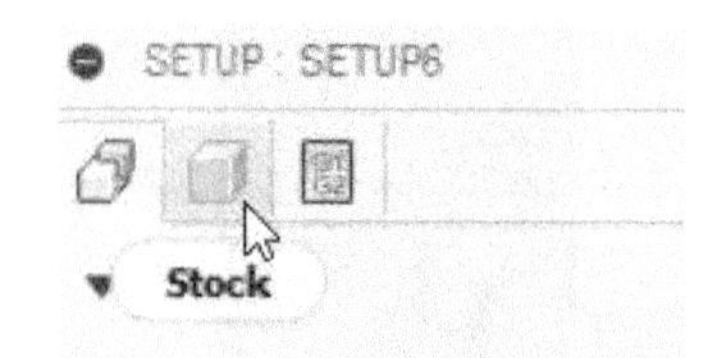

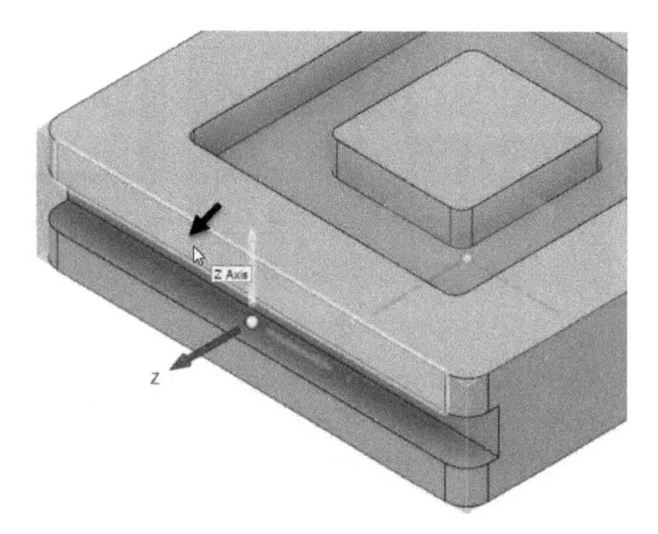

6. Select the front horizontal edge to define the X axis.
7. Leave the WCS at the default location. Next, click **OK** on the **Setup** dialog.
8. Double-click on the name of the **Setup 2**. Next, select all the letters of **Setup 2**.
9. Type **Slot** and then click in the graphics window.

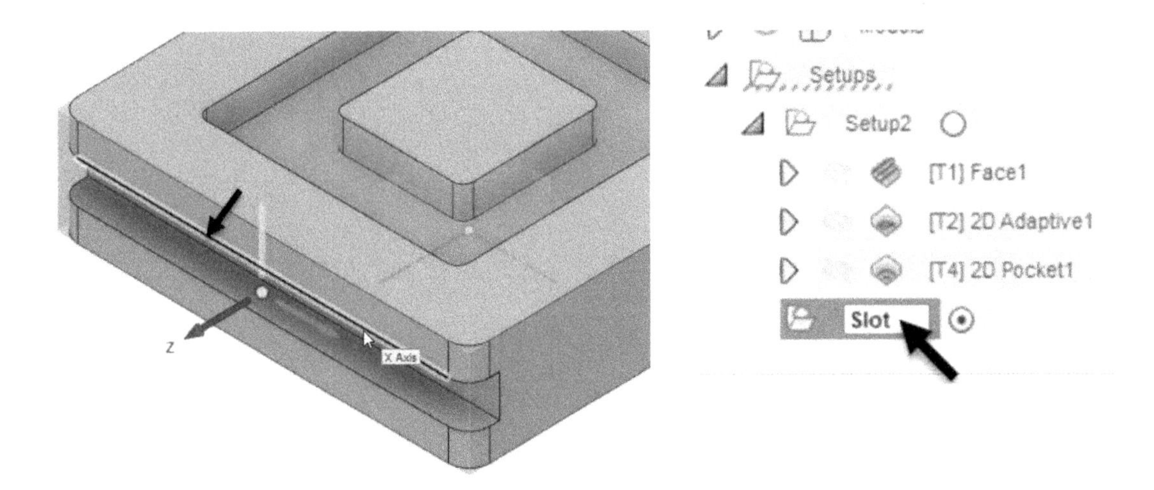

10. On the Toolbar, click **Milling > 2D > 2D Adaptive Clearing**.
11. On the **2D Adaptive** dialog, click the **Select** button next to the **Tool** option.
12. On the **Select Tool** window, select the **Tutorial – Inch** folder from the **Fusion 360 Library**.
13. Select the **4-Ø1/4" L1" (flat end Mill)** tool from the tool list.
14. Click the **Select** button located at the bottom-right corner.
15. Click the **Geometry** tab on the **2D Adaptive** dialog. Next, click on the floor face of the slot, as shown.

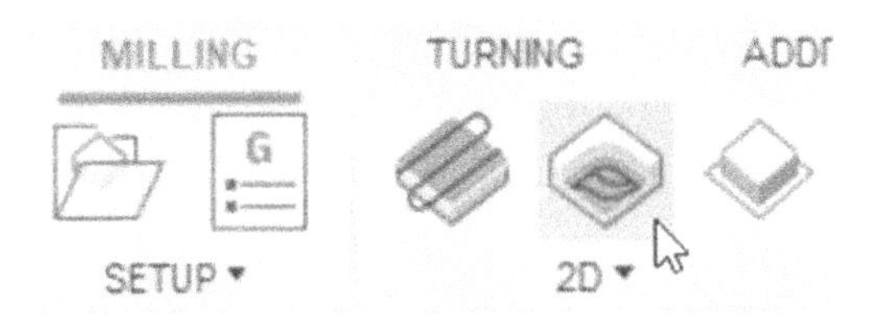

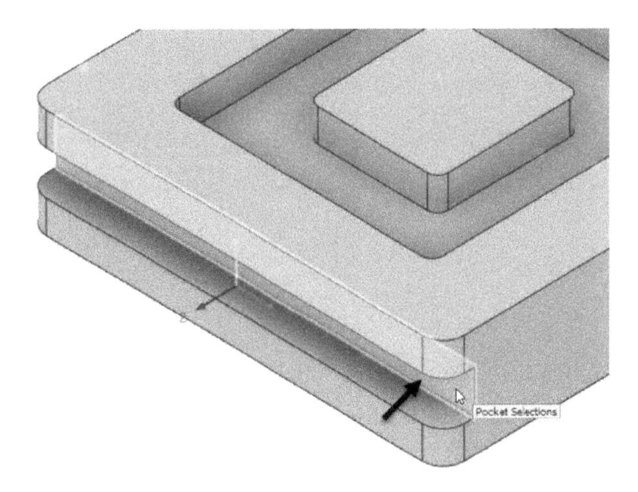

16. Click the **Heights** tab on the **2D Adaptive** dialog.
17. In the **Top Height** section, select **From > Model top**. Next, leave the **Offset** value to **0**.
18. In the **Bottom Height** section, select **From > Selected contours(s)**. Next, leave the **Offset** value to **0**.
19. Click **OK** on the **2D Adaptive** dialog to create the adaptive clearing milling operation.
20. In the Browser, right-click on the **2D Adaptive** operation, and then select **Machining Time**.

Notice that the machining time is displayed as 0:01:20.

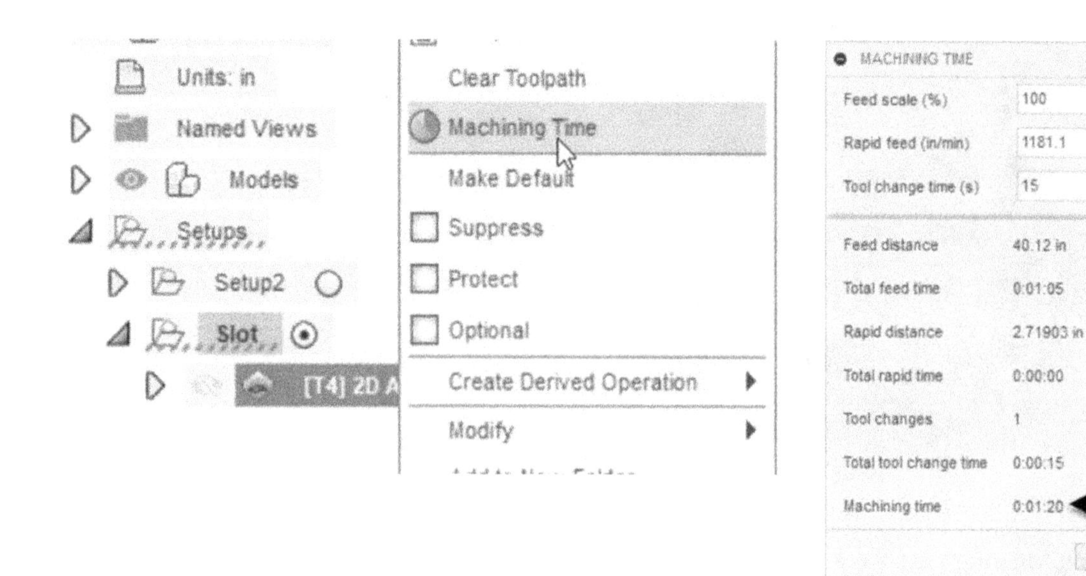

21. Click **Close** on the **Machining Time** dialog.
22. Click **2D Contour** on the **2D** panel.

Notice that the **4-Ø1/4" L1" (flat end Mill)** tool is already selected.

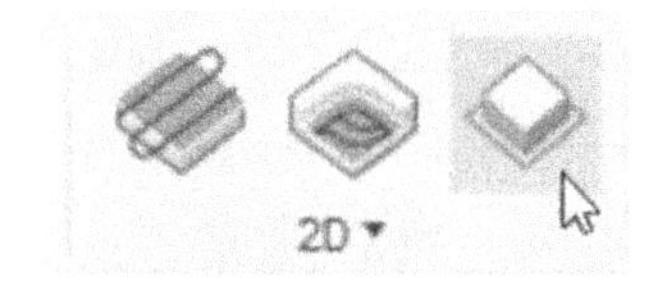

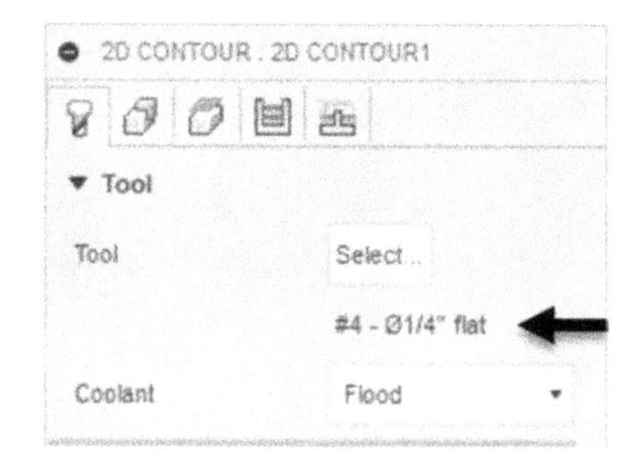

23. Click the **Geometry** tab on the **2D Contour** dialog. Next, click on the edge of the slot, as shown.
24. Rotate the model and select the inner top edge of the slot, as shown.

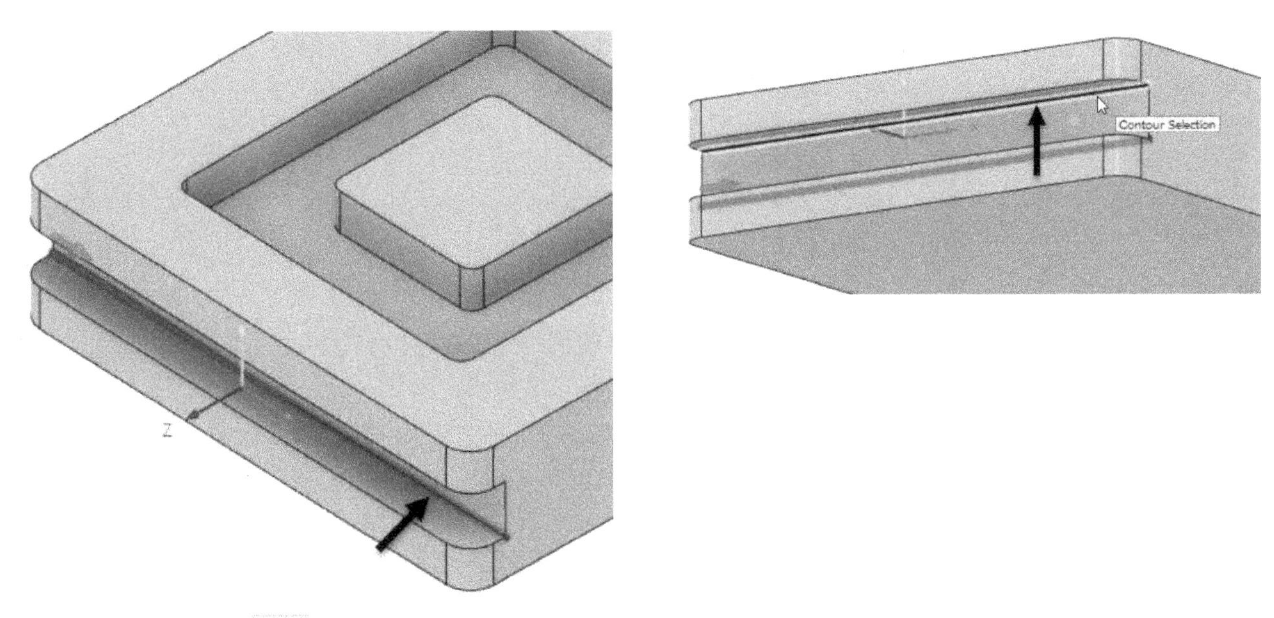

25. Click the **Heights** tab on the **2D Contour** dialog.
26. In the **Top Height** section, select **From > Model top**. The tool starts cutting from the top face of the model.
27. In the **Bottom Height** section, select **From > Selected contours(s)**.
28. Click the **Passes** tab on the **Face** dialog.
29. Click **OK** on the **2D Contour** dialog to clean up the wall of the slot.
30. Likewise, create the 2D Adaptive and 2D Contour operations on the other slot.

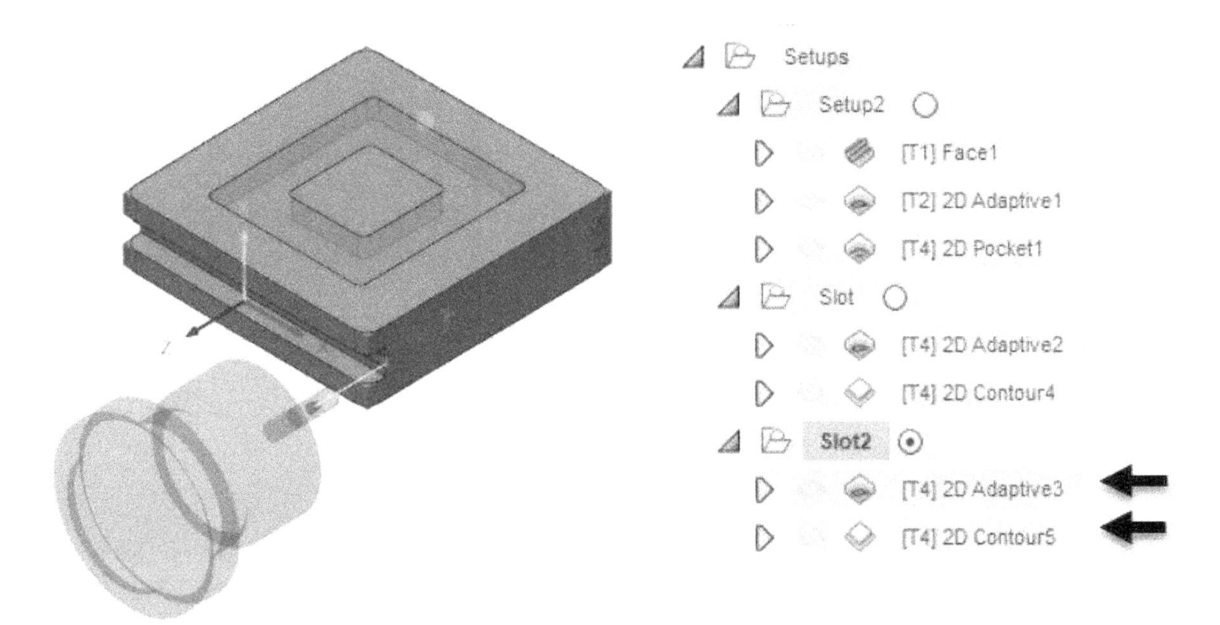

31. Save and close the file.

Tutorial 6

Creating the Boring Operation

1. Download the **Tutorial 6** file, and then upload it to Fusion 360. Next, open the uploaded file.
2. On the Toolbar, click **Change Workspace** drop-down > **Manufacture**.
3. On the Toolbar, click **Milling > Drilling > Drill**.
4. On the **Drill** dialog, click the **Select** button next to the **Tool** option.
5. On the **Select Tool** window, select the **All** > **Local > Library** folder.

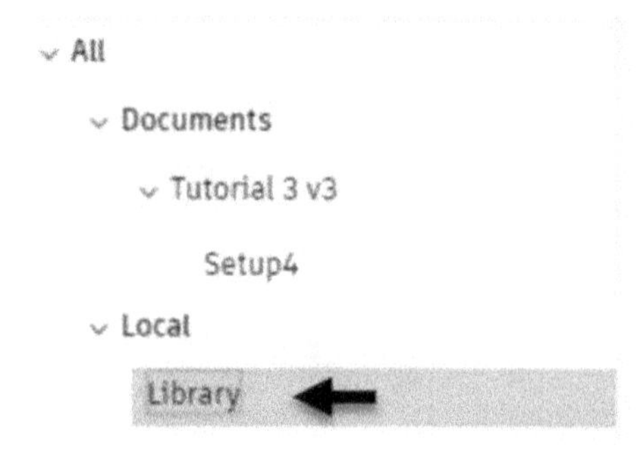

6. Click the **New Tool** icon.
7. On the **New Tool** window, select **Hole making > boring bar**.

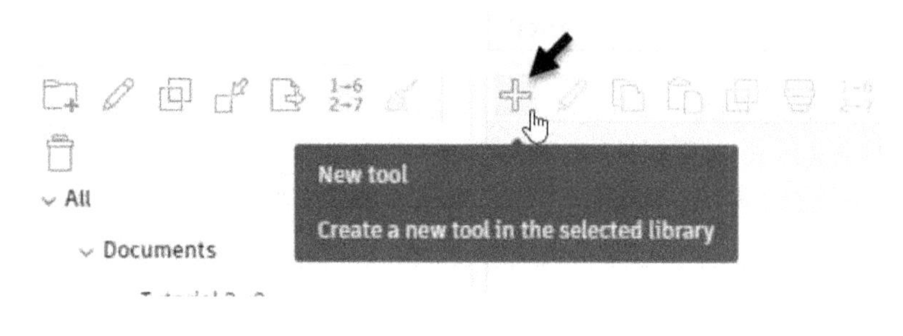

8. Click the **General** tab and enter **BT** in the **Description** box.
9. Click the **Cutter** tab. Next, enter the following values.

Number of flutes: 1
Diameter: 2
Shaft diameter: 2
Overall length: 14
Length below holder: 10
Shoulder length: 6
Flute length: 4

10. Click the **Cutting Data** tab and enter **1500 rpm** in the **Spindle Speed** box. Enter the remaining values as per the manufacturer's recommended speeds and feeds for the tool.
11. Click the **Accept** button. Next, click the **Select** button located at the bottom-right corner.

12. Click the **Geometry** tab on the **Drill** dialog. Next, select the cylindrical face of the hole.
13. Click the **Heights** tab on the **Drill** dialog.
14. Under the **Bottom Height** section, select **From > Hole bottom.** Type **-0.1** in the **Offset** box.
15. Click the **Cycle** tab on the **Drill** dialog. Select **Cycle Type > Boring- dwell and feed out**.
16. Click **OK** on the **Drill** dialog.

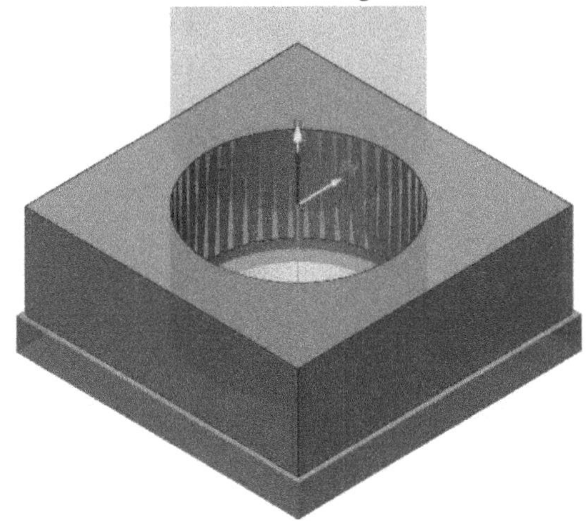

17. Save and close the file.

Tutorial 7

Creating the Bore Milling Operation

1. Download the **Tutorial 7** file, and then upload it to Fusion 360. Next, open the uploaded file.
2. On the Toolbar, click **Milling > 2D > Bore**.
3. On the **Bore** dialog, click the **Select** button next to the **Tool** option.
4. On the **Select Tool** window, select the **Sample Tools - Inch** folder from the **Fusion 360 Library**.
5. Select the **Ø5/8" L1.35" (5/8" Flat Endmill)** tool from the tool list.
6. Select the **Stainless Steel-Finishing** from the **Cutting data** section. Next, click the **Select** button.
7. Click the **Geometry** tab on the **Bore** dialog. Next, select the circular face of the large hole, as shown.

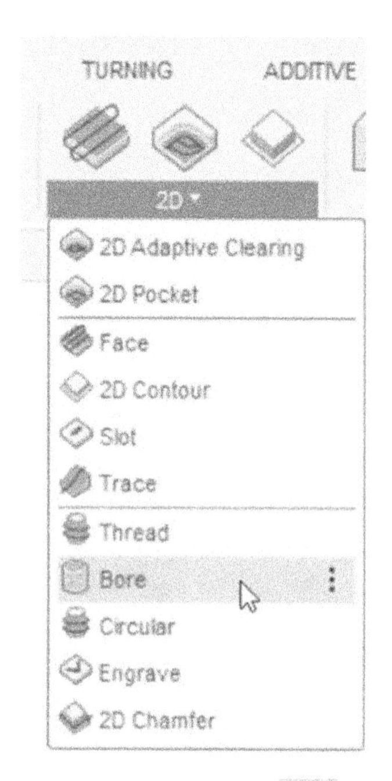

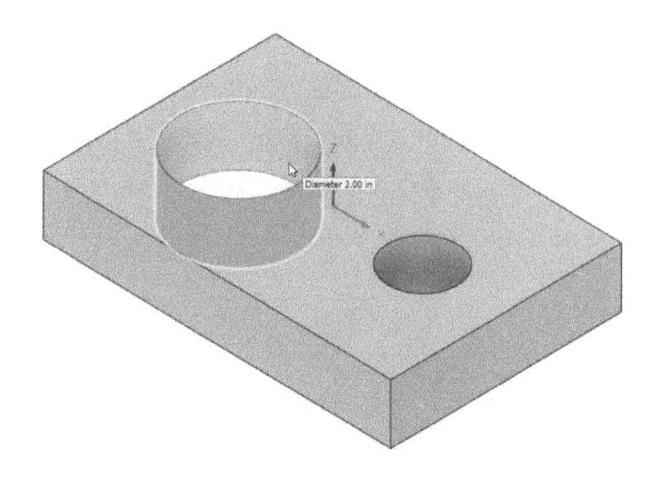

8. Click the **Passes** tab on the **Bore** dialog. Next, select **Compensation Type > Wear**.
9. Type 0.25 in the **Pitch** box. Next, check the **Multiple Passes** option.
10. Type 3 in the **Number of Stepovers** box. Next, type 0.25 in the **Stepover** box.
11. Check the **Finishing Passes** option. Next, select **Direction > Climb**. Next, click **OK** on the **Bore** dialog.

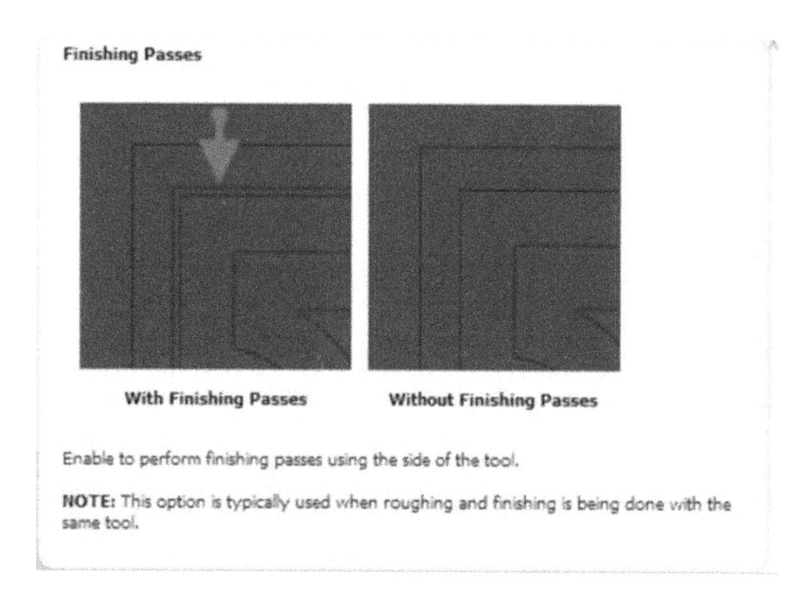

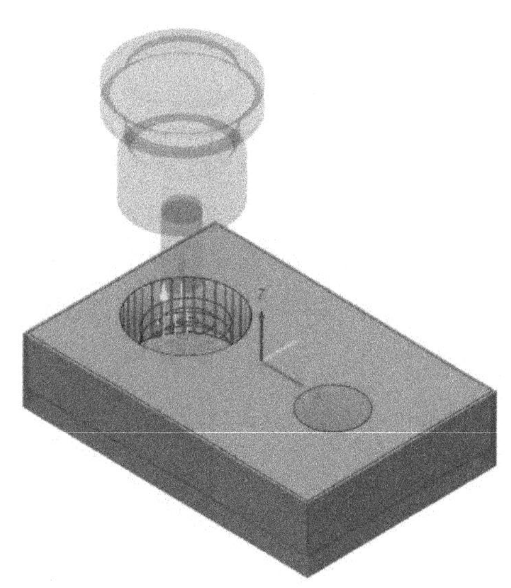

12. In the Browser, right-click on the **Bore** operation, and then select **Duplicate**.
13. In the Browser, right-click on the duplicated **Bore** operation, and then select **Edit**.
14. Click the **Geometry** tab and click the cross mark next to the **Circular Face Selections** box.

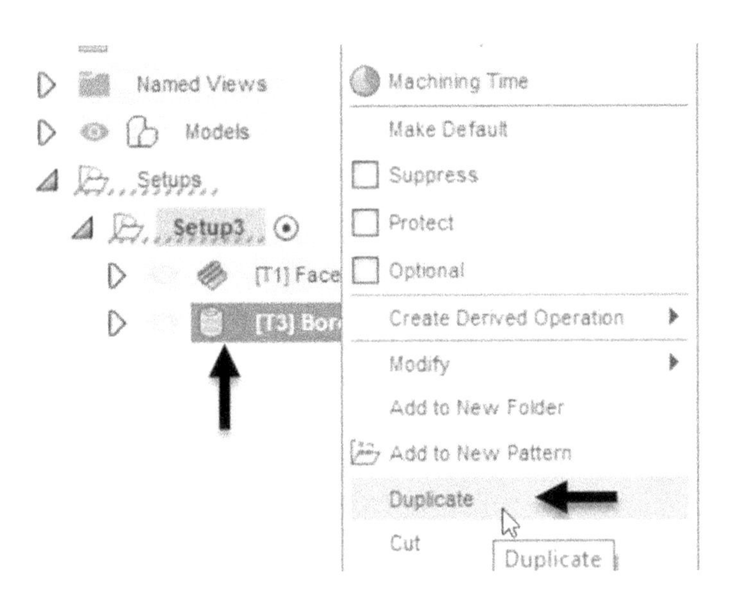

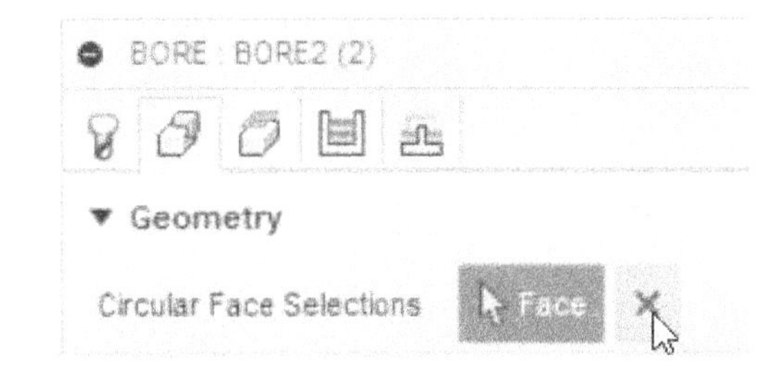

15. Select the small hole.
16. Click the **Passes** tab. Next, uncheck the **Multiple Passes** option.
17. Click **OK**.

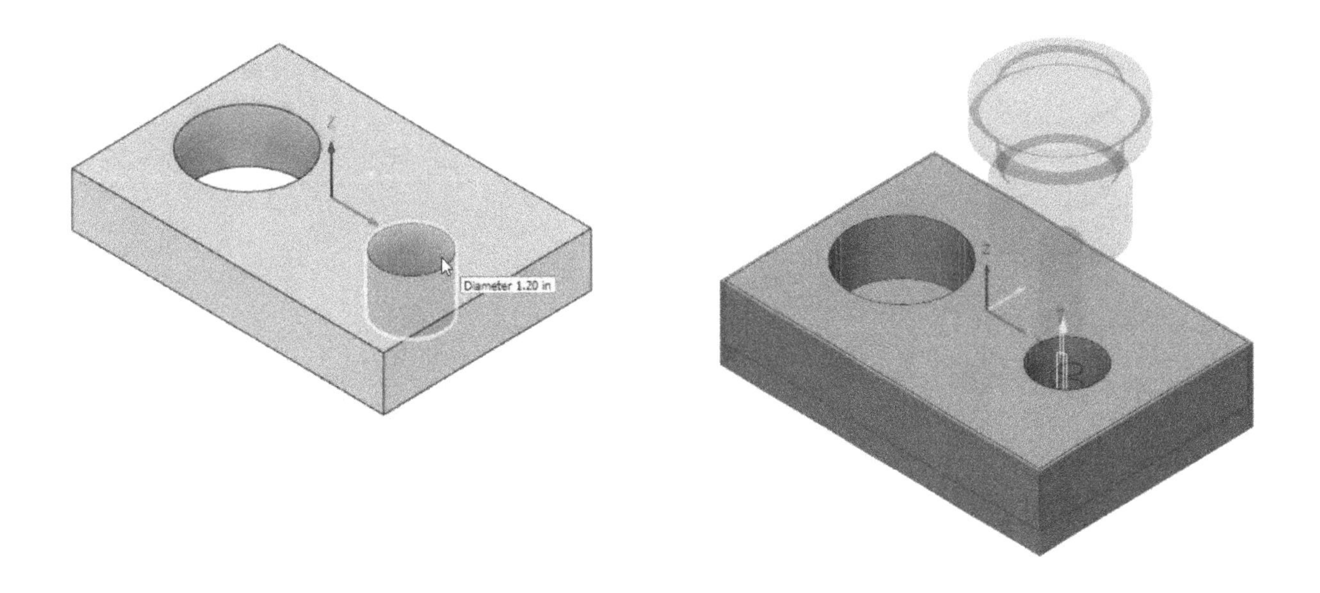

18. Save and close the file.

Tutorial 8

Creating the Trace Operation

1. Download the **Tutorial 8** file, and then upload it to Fusion 360. Next, open the uploaded file.
2. On the Toolbar, click **Milling > 2D > Trace**.

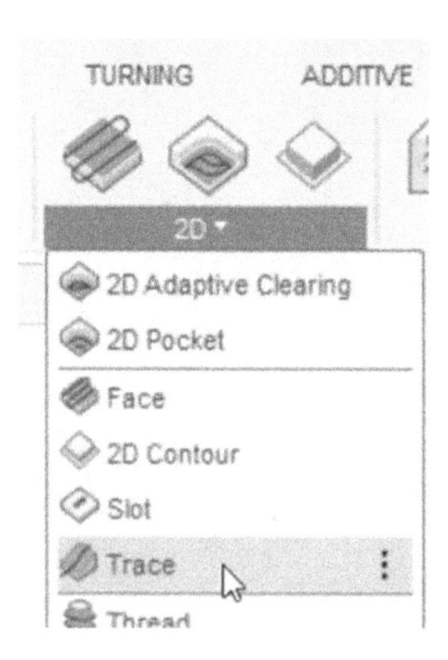

3. On the **Trace** dialog, click the **Select** button next to the **Tool** option.
4. On the **Select Tool** window, select the **Sample Tools – Inch** folder from the **Fusion 360 Library**.
5. Select the **2-Ø1/4" L0.85" (1/4" Spot Drill)** tool from the tool list.
6. Select the **Stainless Steel-Finishing** from the **Cutting data** section. Next, click the **Select** button.
7. Click the **Geometry** tab on the **Trace** dialog.
8. Select the edge loop of the bottom face of the cut feature, as shown.
9. Click the **Passes** tab on the **Bore** dialog. Next, select **Sideways Compensation > Center**.
10. Uncheck the **Chamfer** option. Next, click **OK**.

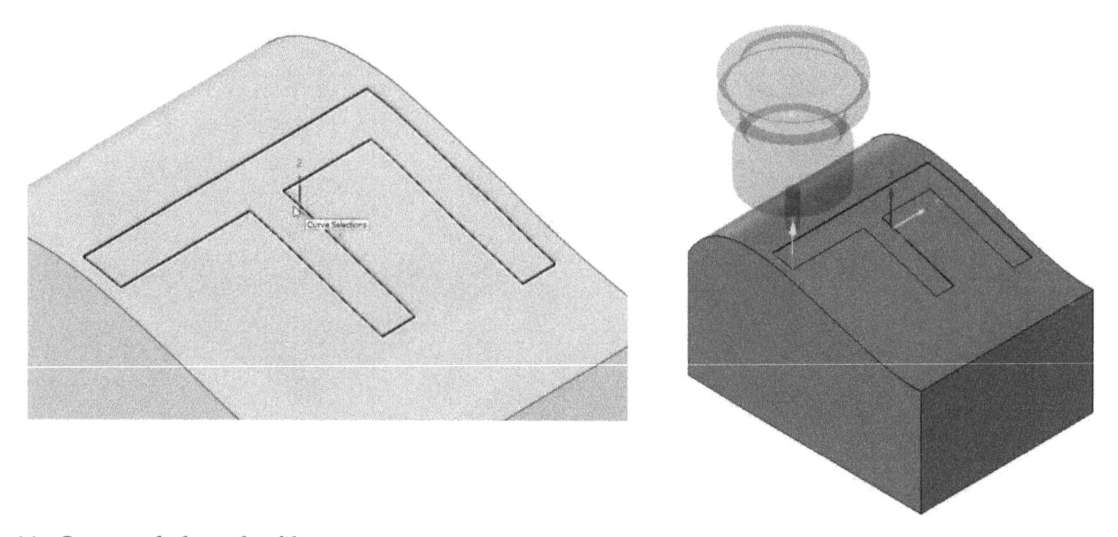

11. Save and close the file.

Tutorial 9

Creating the Engrave Operation

1. Download the **Tutorial 9** file, and then upload it to Fusion 360. Next, open the uploaded file.
2. On the Toolbar, click **Milling > 2D > Engrave**.
3. On the **Engrave** dialog, click the **Select** button next to the **Tool** option.

4. On the **Select Tool** window, select the **All** > **Local > Library** folder. Next, click the **New Tool** icon.
5. On the **New Tool** window, select **Milling > chamfer mill.**

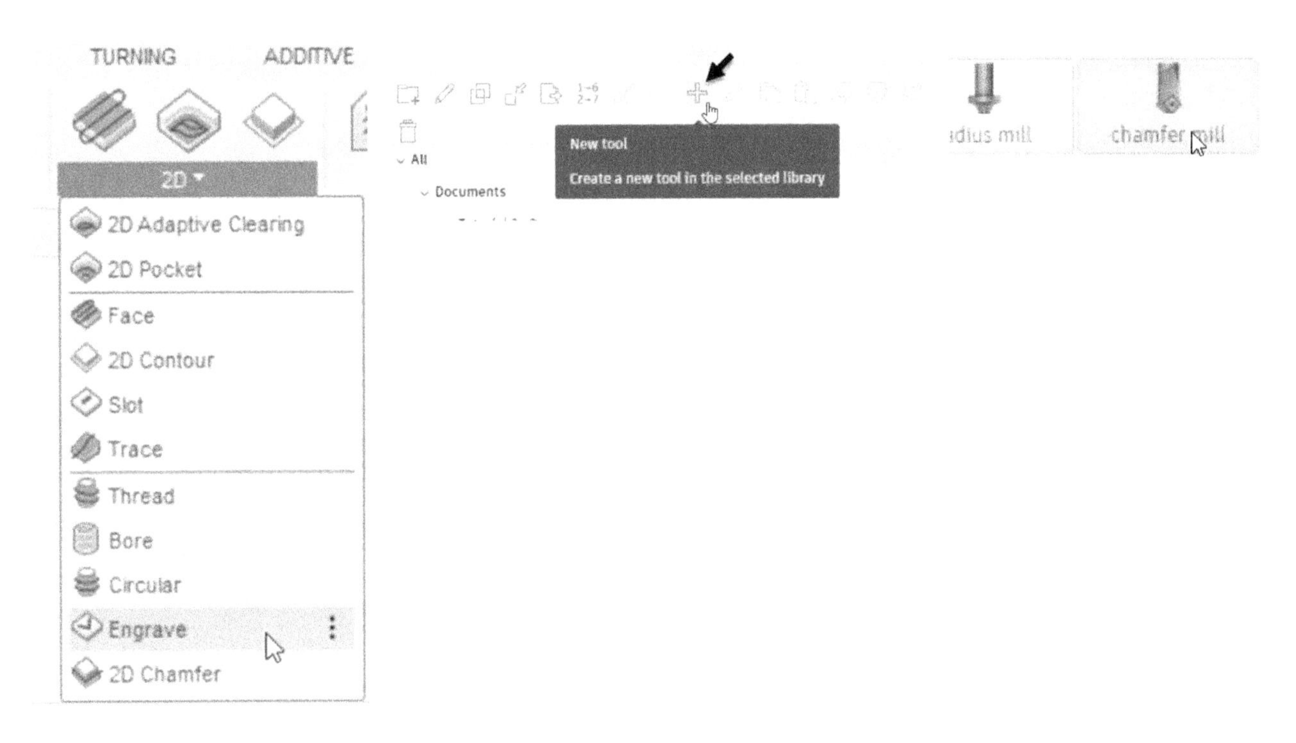

6. On the **General** tab, type **V CUTTER 1/4" SHANK AND BIT END**.
7. Click the **Cutter** tab and specify the following values.

Number of flutes: 2

Material: Carbide

Clockwise spindle rotation: Checked

Diameter: 0.5

Shaft Diameter: 0.25

Tip Diameter: 0

Overall Length: 1.5

Length below holder: 1

Shoulder Length: 0.5

Flute Length: 0.5

Taper angle: 45 degrees

8. Click **Accept** and **Select**.
9. Click the **Geometry** tab on the **Trace** dialog. Next, select the top edges loop of the cut features, as shown.
10. Click the **Heights** tab. Next, type **-0.1** in the **Offset** box available in the **Bottom Height** section.
11. Click **OK**.
12. In the Browser, right-click on the **Engrave** operation, and then select **Simulate**.

13. Close the **Simulate** dialog.
14. Save and close the file.

Tutorial 10

Creating Chamfers using the 2D Contour operation

1. Download the **Tutorial 10** file, and then upload it to Fusion 360. Next, open the uploaded file.
2. On the Toolbar, click **Milling > 2D > 2D Contour**.
3. On the **2D Contours** dialog, click the **Select** button next to the **Tool** option.
4. On the **Select Tool** window, select the **All** folder.
5. Click the **Filters** tab on the top-right corner. Next, select **Tool Category > Milling**.
6. Select **Type > chamfer mill**. Next, select the **19 - Ø1/2" L1.25" (chamfer mill)** tool from the tool list.
7. Click the **Select** button located at the bottom-right corner. Next, select the top edges, as shown.

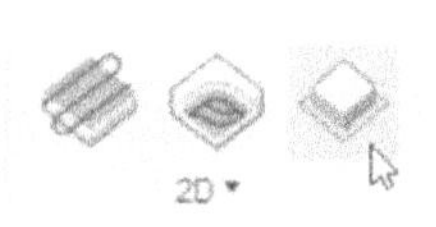

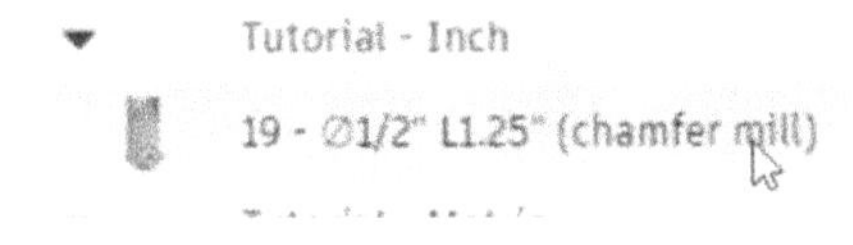

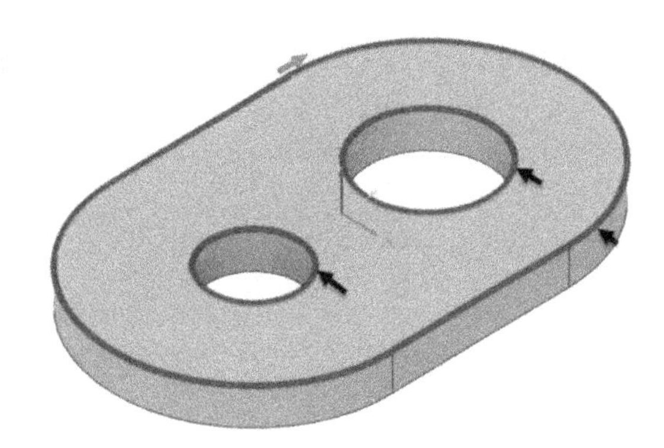

8. Click the **Heights** tab.
9. On the **Top Height** section, select **From > Model top**.
10. Click the **Passes** tab. Notice that the **Chamfer** option is selected.
11. Type **0.02** in the **Chamfer width** box.
12. Type **0.04** in the **Chamfer Tip Offset** box. Next, click **OK** on the **2D Contours** dialog.

Chamfer Tip Offset

This is added to the toolpath depth, while keeping the tool in contact with the selected edge by adjusting the toolpath radial offset.

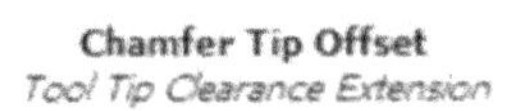

NOTE: Allows you to cut past the bottom of the Chamfer edge. Keeps you from cutting with the bottom tool edge.

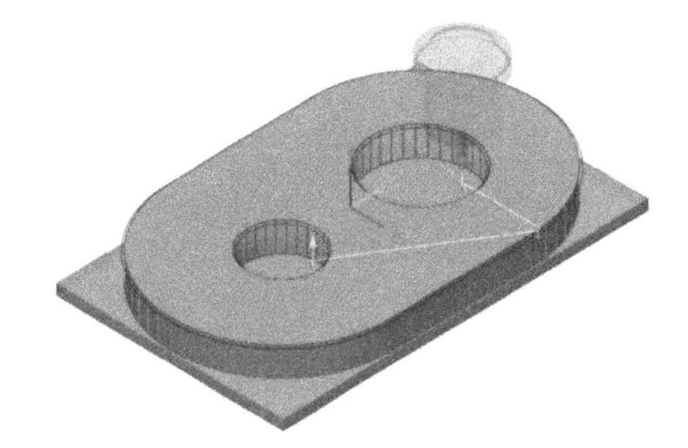

Tutorial 11

Creating Chamfers using the 2D Chamfer operation

This operation enables you to break the sharp edges of the part. It is similar to the 2D Contour operation.

1. Download the **Tutorial 11** file and then upload it to Fusion 360. Next, open the uploaded file.
2. On the Toolbar, click **Milling > 2D > 2D Chamfer**.
3. On the **2D Chamfer** dialog, click the **Select** button next to the **Tool** option.
4. On the **Select Tool** window, select the **All** folder. Next, click the **Filters** tab on the top-right corner.

5. Select **Tool Category > Milling**. Next, select **Type > chamfer mill.**
6. Select the **19 - Ø1/2" L1.25" (chamfer mill)** tool from the tool list.
7. Click the **Select** button located at the bottom-right corner.
8. Select the edge loop of the model, as shown.

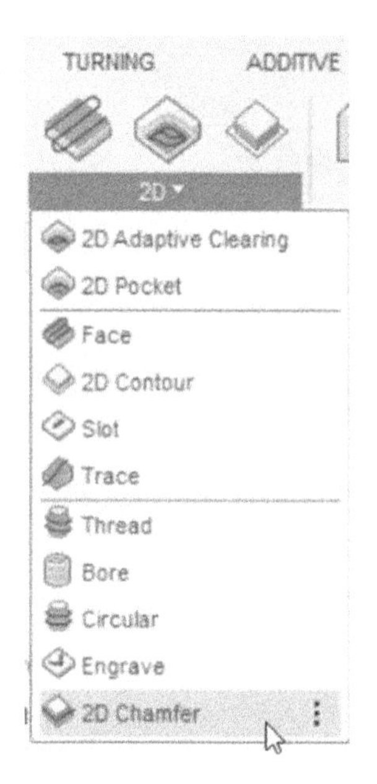

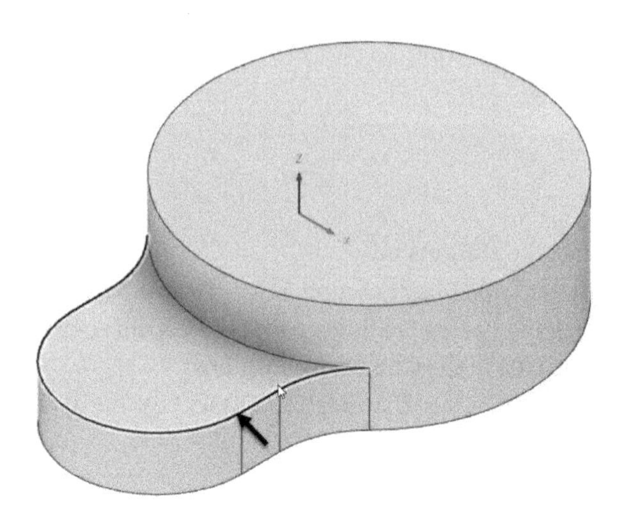

9. Click the **Passes** tab. In the **Chamfer** section, type 0.1 in the **Chamfer width** box.
10. Type **0.05** in the **Chamfer Tip Offset** box. Next, type **0.005** in the **Chamfer Clearance** box.

    The chamfer clearance is the clearance value between the tool and the rest of the model.

11. Click **OK**.

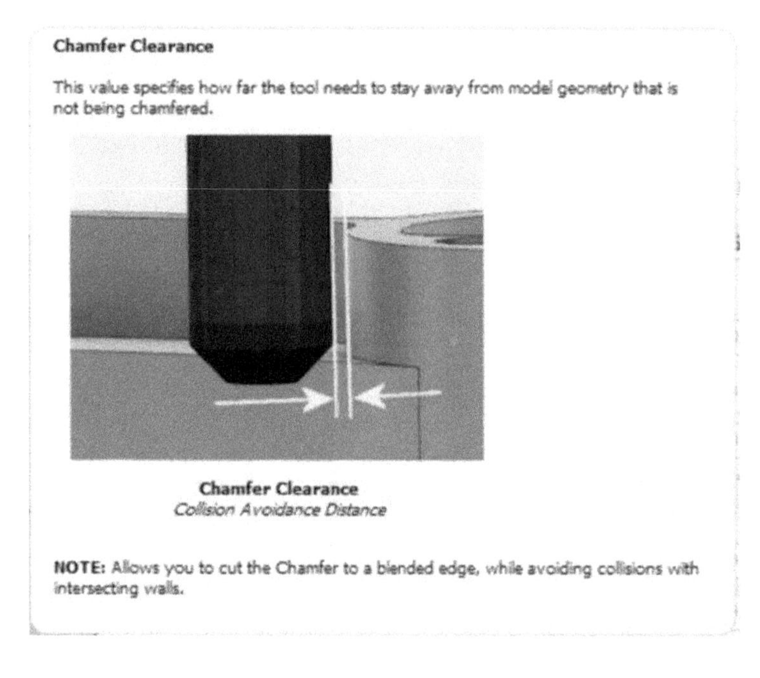

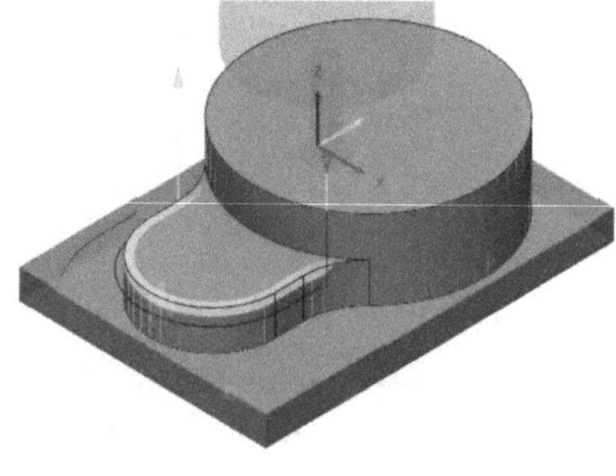

12. Save and close the file.

Tutorial 12

Creating the 3D Adaptive Clearing Operation

1. Download the **Tutorial 12** file, and then upload it to Fusion 360. Next, open the uploaded file.
2. On the Toolbar, click **Milling > Setup > Setup**.
3. On the **Setup** dialog, click the **Select** button next to the **Machine** option.
4. Select the **Autodesk Generic 3-axis** machine from the **Machine Library** dialog.
5. Click the **Select** button. Next, select **Orientation > Select Z axis/plane & X axis**.
6. Select the Y-axis of the coordinate system to define the Z axis.
7. Leave the WCS at the default location.
8. Click the **Stock** tab on the **Setup** dialog. Next, select **Mode > Relative size box**.
9. Set **Stock Side Offset** to **0**. Next, set **Stock Top Offset** to **0.04**.
10. Set **Stock Bottom Offset** to **0**. Next, click **OK** on the **Setup** dialog.
11. On the Toolbar, click **Milling > 3D > Adaptive Clearing**.
12. On the **Adaptive** dialog, click the **Select** button next to the **Tool** option.
13. On the **Select Tool** window, select the **Sample Tools – Inch** folder from the **Fusion 360 Library**.
14. Select the **Ø3/8" L0.975" (3/8" Bullnose Endmill)** tool from the tool list.
15. Select the **Aluminum – Roughing** option from the **Cutting Data** section.
16. Click the **Select** button located at the bottom-right corner.
17. Click the **Geometry** tab and select the rectangular edge of the model, as shown.

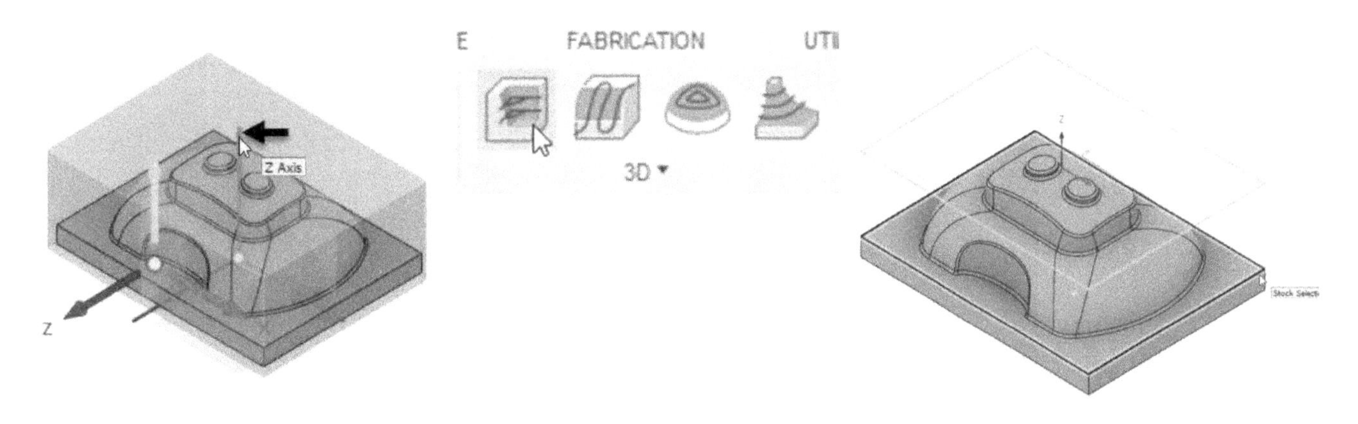

18. Under the **Rest Machining** section, select **Adjustments > Use as computed**.
19. Click the **Passes** tab on the dialog. Next, right-click in the **Optimal Load** box and select **Edit Expression**.
20. Change the expression to **tool_diameter*0.2**.
21. Click **OK** on the **Expression** dialog.
22. Type **0.0375** in the **Maximum Roughing Stepdown** box. Next, click **OK**.
23. Select the **Adaptive** operation from the Browser.
24. On the Toolbar, click **Milling > Actions > Simulate**.
25. On the **Simulate** dialog, under the **Stock** section, select **Mode > Standard**.
26. Select **Colorization > Comparison**. Next, click the **Start the simulation** play button.

Notice that the machined part is displayed in blue color. The blue color indicates leftover material. You need to perform more milling operations to remove the blue material.

27. Click **Close** on the **Simulate** dialog.
28. On the Toolbar, click **Milling > 3D > Adaptive Clearing**.
29. On the **Adaptive** dialog, click the **Select** button next to the **Tool** option.
30. On the **Select Tool** window, select the **Sample Tools - Inch** folder from the **Fusion 360 Library**.
31. Select the **Ø1/8" L0.6" (1/8" Bullnose Endmill)** tool from the tool list.
32. Select the **Aluminum - Roughing** option from the **Cutting Data** section.
33. Click the **Select** button located at the bottom-right corner.
34. Click the **Geometry** tab.
35. Under the **Rest Machining** section, select **Source > From previous operations(s)**.
36. Select **Adjustments > Use a computed**.
37. Click the **Passes** tab on the dialog. Next, right-click in the **Optimal Load** box and select **Edit Expression**.
38. Change the expression to **tool_diameter*0.2**. Next, click **OK** on the **Expression** dialog.
39. Type **0.03** in the **Maximum Roughing Stepdown** box. Next, click **OK**.
40. Right click on the second **Adaptive** operation and select **Simulate**.

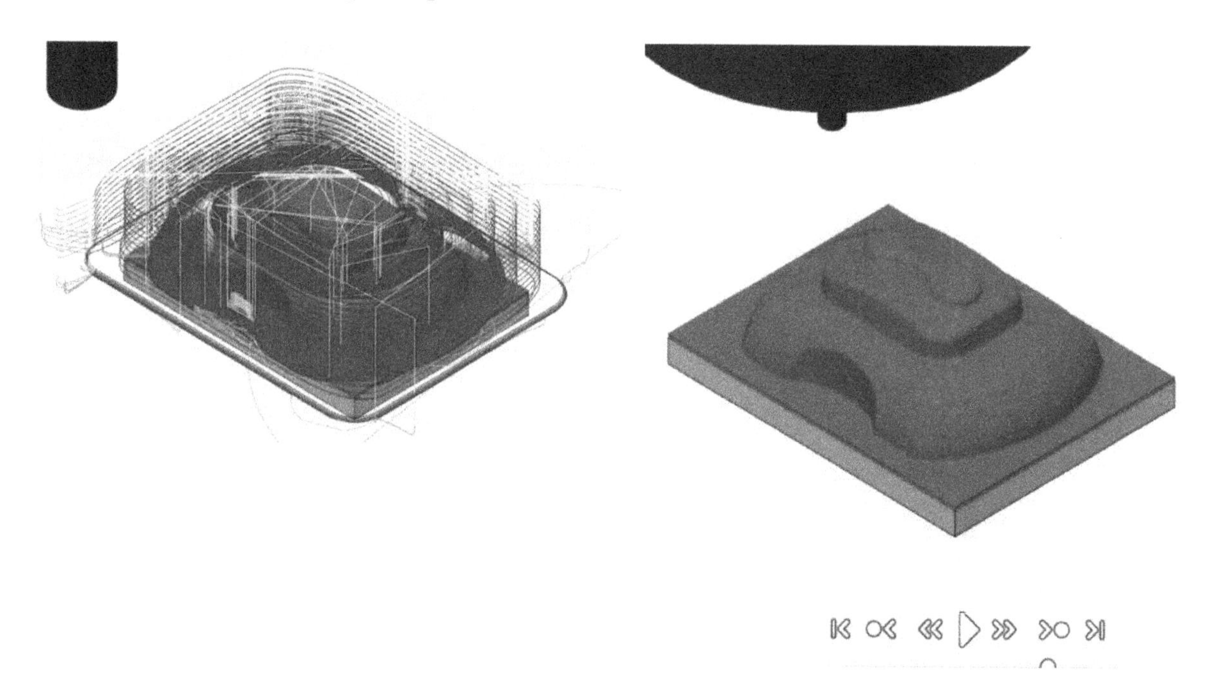

41. Close the **Simulate** dialog.

Creating the 3D Contour Operation

1. On the Toolbar, click **Milling > 3D > Contour**.

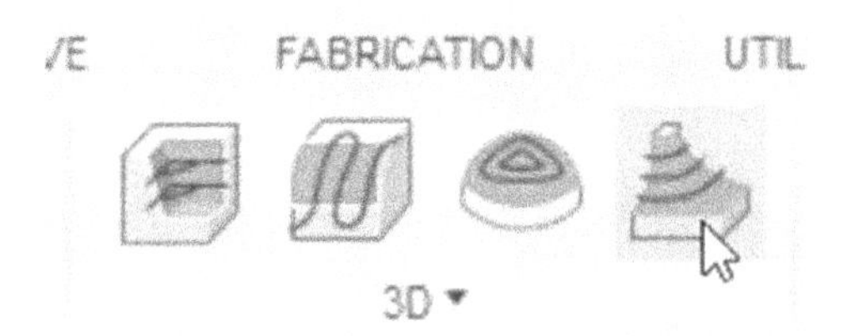

2. Click the **Select** button next to the **Tool** option.
3. Select the **Ø1/8" L0.6" (1/8" Bullnose Endmill)** tool from the tool list.
4. Select the **Aluminum - Finishing** option from the **Cutting Data** section. Next, click the **Select** button.
5. Click the **Geometry** tab. Next, check the **Slope** option
6. Under the **Slope** section, type **30** in the **From Slope Angle** box. Next, type **90** in the **To Slope Angle** box.
7. Click the **Heights** tab. Next, select **From > Selection** from the **Bottom Height** section.
8. Select the flat face of the model.
9. Click the **Passes** tab. Next, type **0.01** in the **Maximum Stepdown** box and click **OK**.
10. Right click on the second **Contour** operation and select **Simulate**.

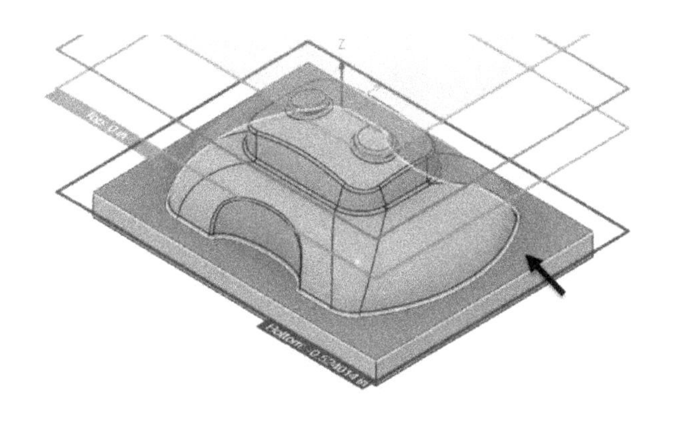

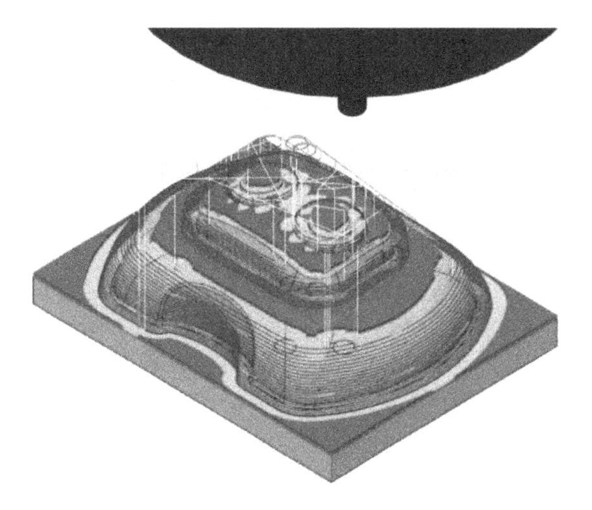

11. Close the **Simulate** dialog.

Creating the Parallel Operation

1. On the Toolbar, click **Milling > 3D > Parallel**.

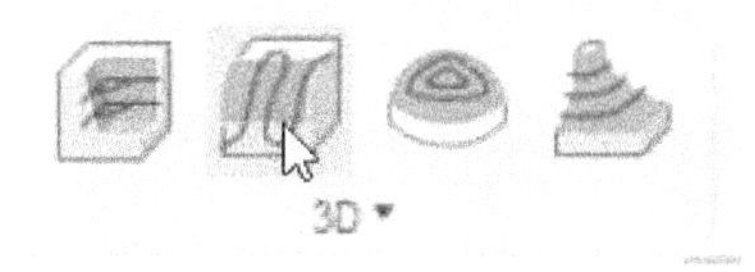

2. Click the **Geometry** tab. Next, check the **Slope** option
3. Under the **Slope** section, type **0** in the **From Slope Angle** box. Next, type **35** in the **To Slope Angle** box.
4. Check the **Avoid/Touch Surfaces** option. Next, select the flat face, as shown.
5. Click the **Passes** tab. Next, type **45** in the **Pass Direction** box.
6. Type **0.01** in the **Stepover** box, and then click **OK**.
7. Right click on the second **Parallel** operation and select **Simulate**.
8. Click the **Start the simulation** button.

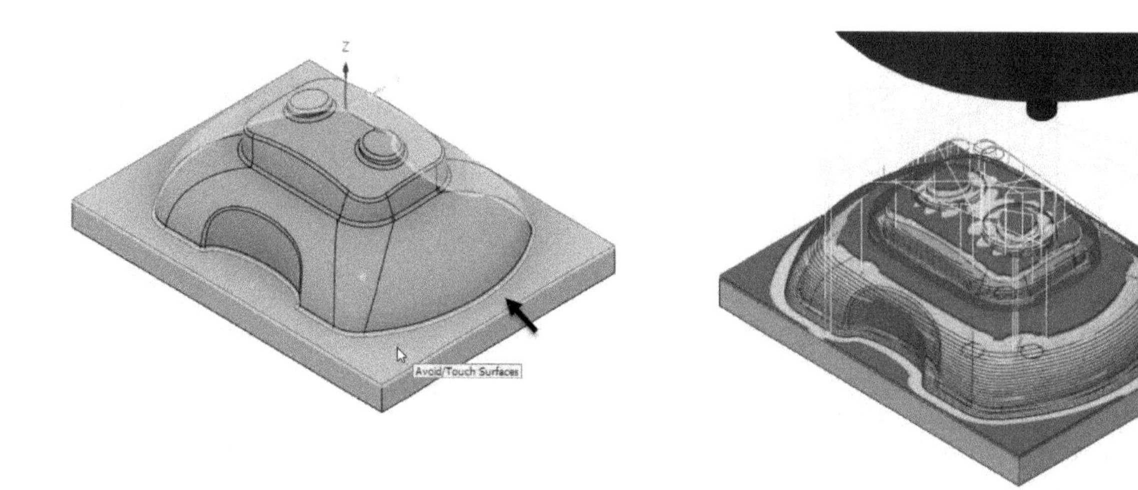

9. Close the **Simulate** dialog.
10. Save and close the design file.

Tutorial 13

Creating the Spiral Operation

1. Download the **Tutorial 13** file, and then upload it to Fusion 360. Next, open the uploaded file.
2. On the Toolbar, click **Milling > 3D > Spiral**. Next, click the **Select** button next to the **Tool** option.
3. Select the **Ø5/8" L1.35" (5/8" Bullnose Endmill)** tool from the tool list.
4. Select the **Aluminum - Finishing** option from the **Cutting Data** section. Next, click the **Select** button.
5. Click the **Geometry** tab. Next, check the **Rest Machining** option.
6. Under the **Rest Machining** section, select **Source > From previous operations(s)**.
7. Select **Adjustments > Use as computed**.
8. Click the **Passes** tab on the dialog. Next, type **0.0625** in the **Stepover** box, and then click **OK**.
9. Simulate the operation and notice that the face is not machined fully.

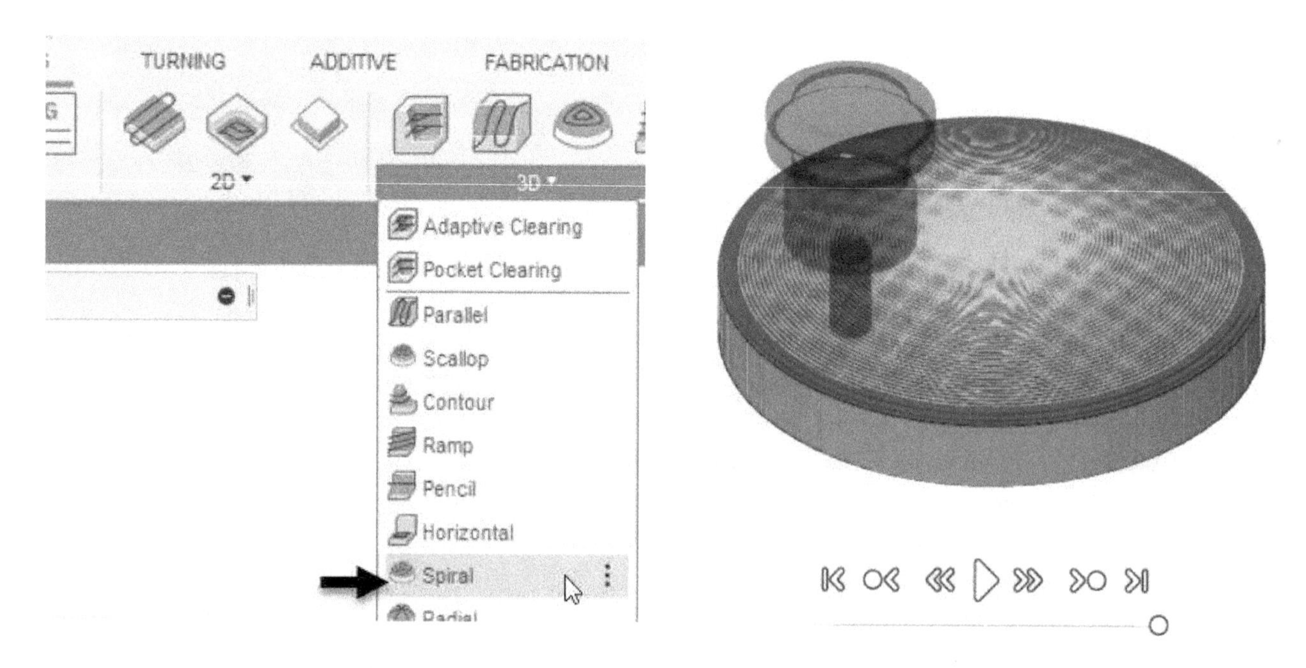

10. Close the **Simulate** dialog.
11. In the Browser, right-click on the **Spiral** operation and select **Edit**.
12. Click the **Geometry** tab. Next, type **0.5** in the **Additional Offset** box, and then click **OK**.
13. Simulate the operation and notice that the face is machined completely.

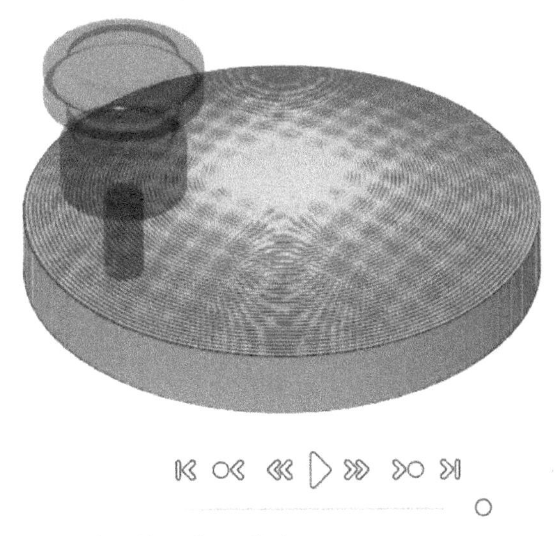

14. Close the **Simulate** dialog.

Creating the Radial Operation

1. On the Toolbar, click **Milling > 3D > Radial**.
2. Click the **Geometry** tab. Next, select **Machining Boundary > Selection**.
3. Select the top circular edge of the model, as shown.

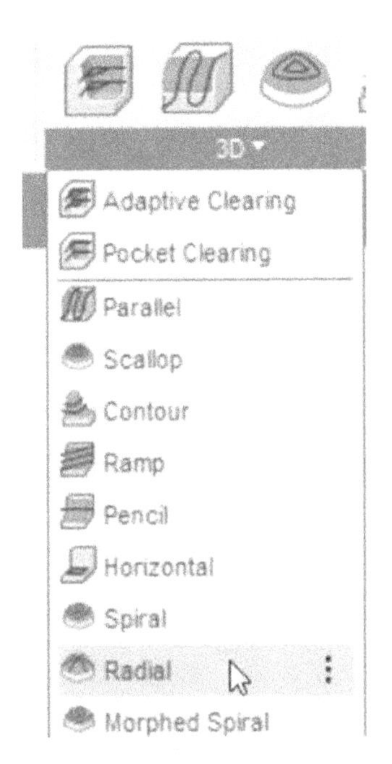

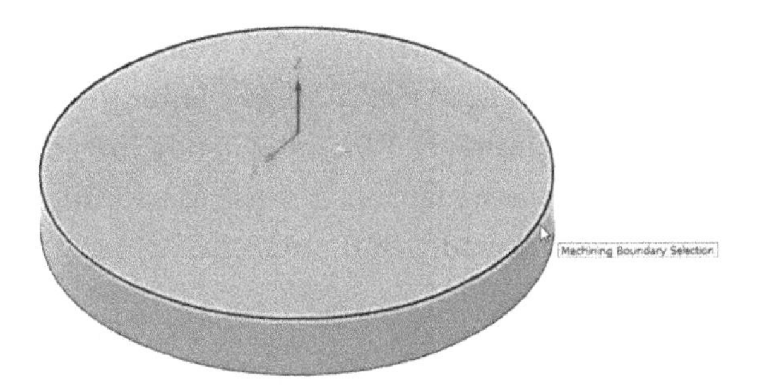

4. Type **0.5** in the **Additional Offset** box. Next, check the **Rest Machining** option.

5. Under the **Rest Machining** section, select **Source > From previous operations(s)**.
6. Select **Adjustments > Use as computed**.
7. Click the **Passes** tab on the dialog. Next, type **1** in the **Angular Step** box, and then click **OK**.
8. Simulate the operation and notice that the face is machined completely.

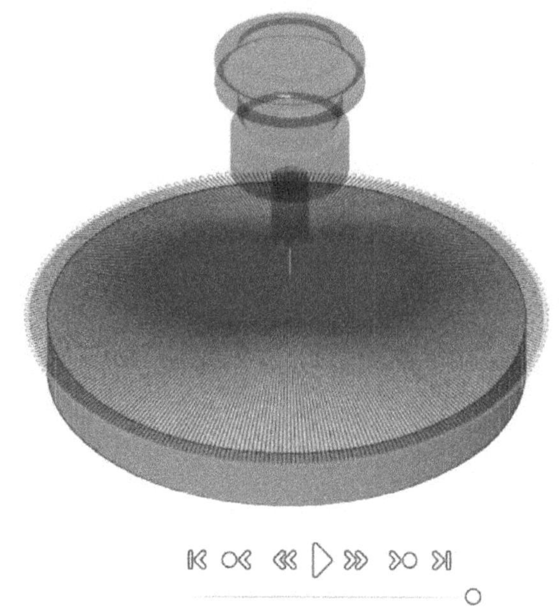

9. Close the **Simulate** dialog.
10. Save and close the design file.

Tutorial 14

Creating the Horizontal Operation

1. Download the **Tutorial 14** file, and then upload it to Fusion 360. Next, open the uploaded file.
2. On the Toolbar, click **Milling > 3D > Horizontal**. Next, click the **Select** button next to the **Tool** option.
3. Select the **Ø1/8" L0.6" (1/8" Bullnose Endmill)** tool from the tool list.
4. Select the **Aluminum - Roughing** option from the **Cutting Data** section. Next, click the **Select** button.
5. Click the **Geometry** tab. Next, select **Machining Boundary > Selection**.
6. Select the circular edge of the model, as shown.

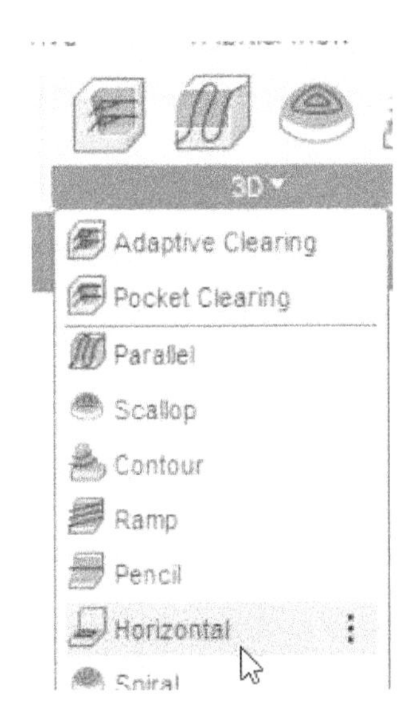

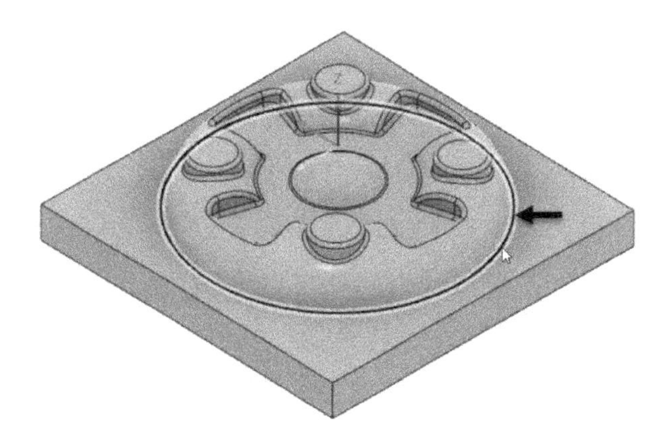

7. Click the **Passes** tab. Next, type **0.01** in the **Smoothing Deviation** box.
8. Type **0.04** in the **Minimum Cutting Radius** box.

Smoothing Deviation

The maximum amount of smoothing applied to the roughing passes. Use this parameter to avoid sharp corners in the toolpath.

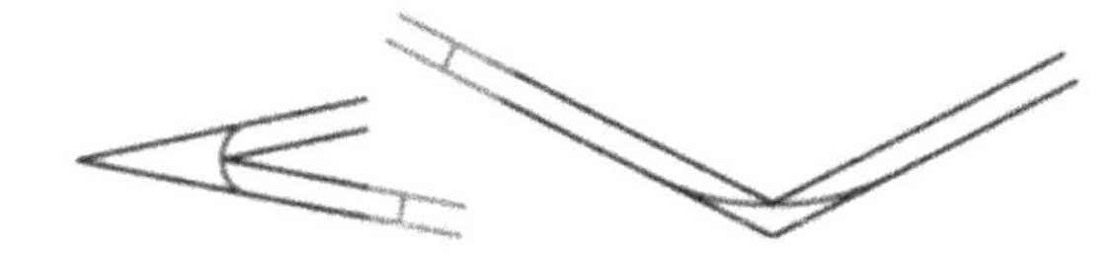

Minimum Cutting Radius

Defines the smallest toolpath radius to generate in a sharp corner. Minimum Cutting Radius creates a blend at all inside sharp corners.

Forcing the tool into a sharp corner, or a corner where the radius is equal to the tool radius, can create chatter and distort the surface finish.

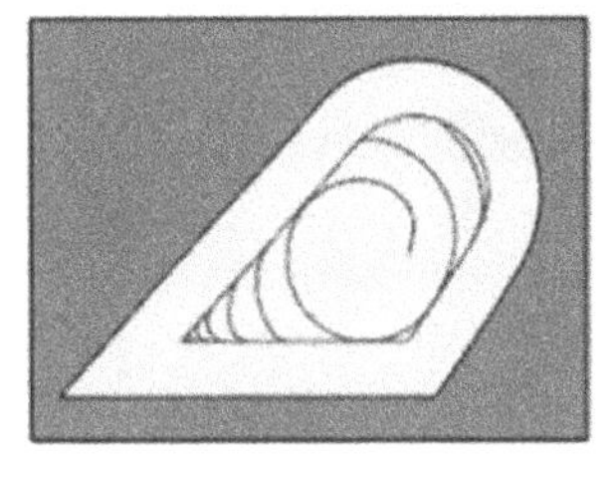

Set to Zero
The toolpath is forced into all inside sharp corners

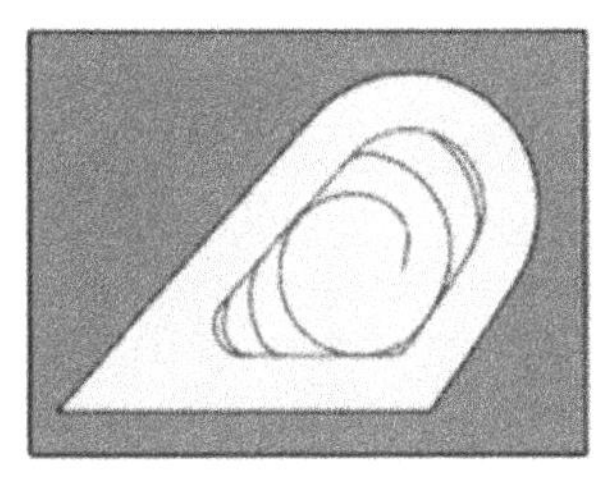

Set to 0.07 in
The toolpath will have a blend of .070 radius in all sharp corners.

NOTE: Setting this parameter leaves more material in internal corners, requiring subsequent rest machining operations with a smaller tool.

9. Check the **Axial Offset Passes** option. Next, type **0.0125** in the **Maximum Stepdown** box.
10. Type **5** in the **Number of Stepdowns** box. Next, check the **Order by Depth** option.

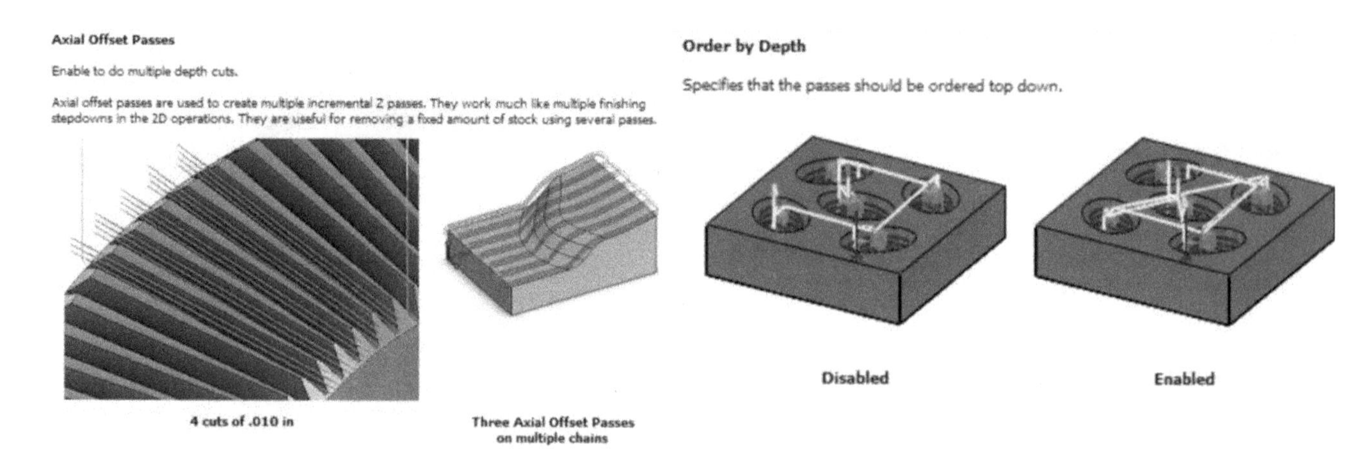

11. Click the **Linking** tab.
12. Under the **Ramp** section, select **Ramp Type > Helix**. Next, click **OK.**
13. Simulate the model and close the **Simulate** dialog.

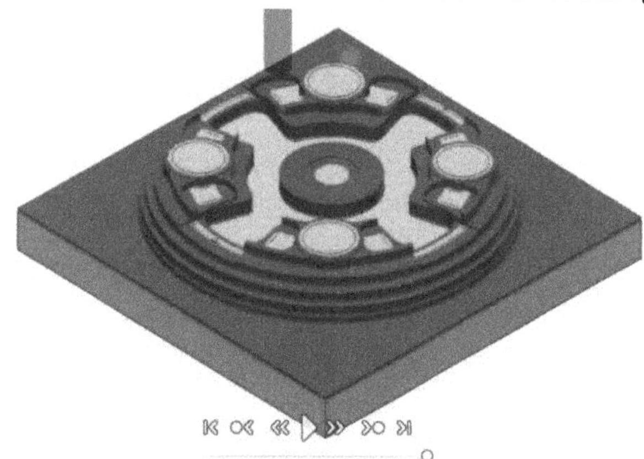

Creating the Scallop Operation

1. On the Toolbar, click **Milling > 3D > Scallop**.

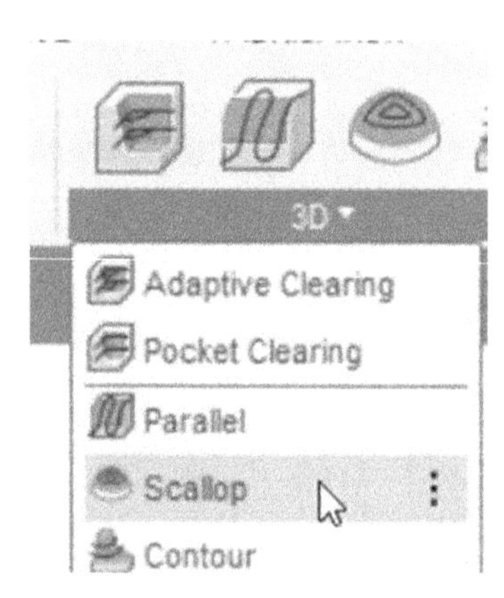

2. Click the **Select** button next to the **Tool** option.
3. Select the **Ø1/8" L0.6" (1/8" Bullnose Endmill)** tool from the tool list.
4. Select the **Aluminum – Finishing** option from the **Cutting Data** section. Next, click the **Select** button.
5. On the **Scallop** dialog, type **4000** in the **Spindle Speed** box. Next, type **20** in the **Cutting Feedrate** box.
6. Click the **Geometry** tab. Next, select **Machining Boundary > Selection**.

7. Select the circular edge of the model, as shown. Next, type **0.1** in the **Additional Offset** box.
8. Check the **Rest Machining** option.
9. Under the **Rest Machining** section, select **Source > From previous operations(s)**.
10. Select **Adjustments > Machine cusps**.
11. Click the **Passes** tab on the dialog. Next, type **0.01** in the **Stepover** box.
12. Set the **Direction** to **Both ways**. Next, check the **Smoothing** option, and then click **OK**.
13. Simulate the operation and notice that the face is machined completely.

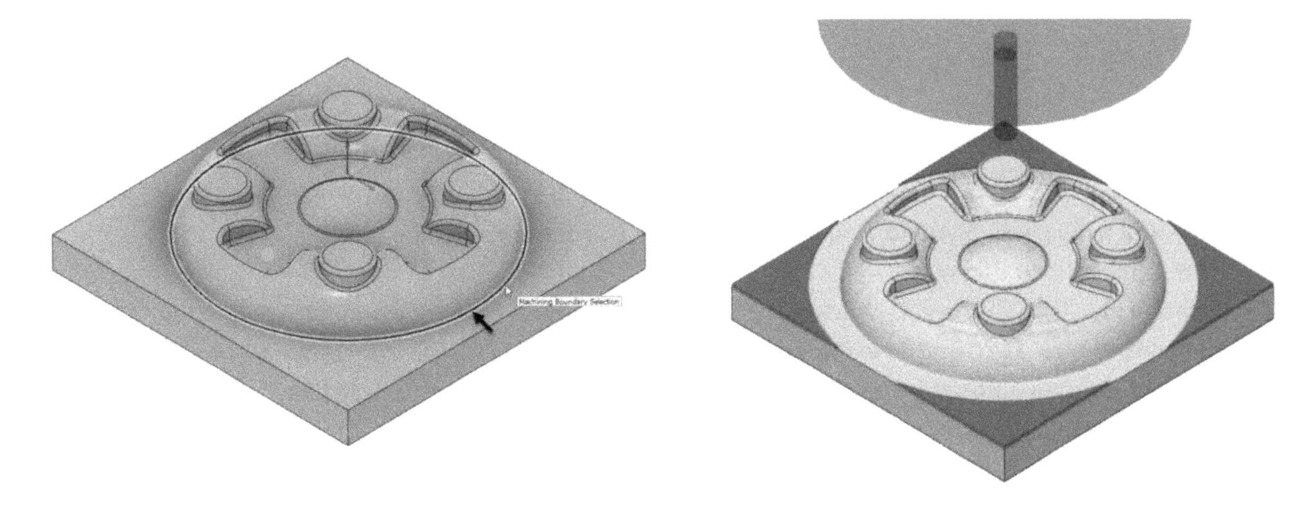

14. Close the **Simulate** dialog.
15. Save and close the design file.

Tutorial 15

Creating the Morph Operation

1. Download the **Tutorial 15** file, and then upload it to Fusion 360. Next, open the uploaded file.
2. On the Toolbar, click **Milling > Setup > Setup.**
3. On the **Setup** dialog, click the **Select** button next to the **Machine** option.
4. On the **Machine Library** dialog, select the **Fusion 360 Library** option from the left side.
5. Select the **Autodesk Generic 3-axis** machine from the list. Next, click the **Select** button.
6. On the **Setup** dialog, select **Operation Type > Milling**.
7. On the **Setup** dialog, select **Orientation > Select Z Axis/plane & X axis**.
8. Select the vertical axis of the coordinate system to define the Z axis.
9. Click the **Box Point** button next to the **Stock Point** option.
10. Select the top left corner point of the stock, as shown.

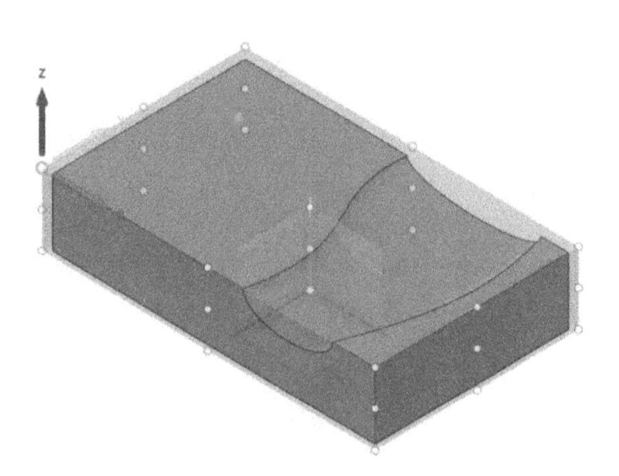

11. Click the **Stock** tab on the **Setup** dialog. Next, select **Mode > From fixed size box**.
12. Click **OK** on the **Setup** dialog.
13. On the Toolbar, click **Milling > 3D > Morph**. Next, click the **Select** button next to the **Tool** option.
11. Select the **Ø1/8" L0.6" (1/8" Bullnose Endmill)** tool from the tool list.
12. Select the **Aluminum - Roughing** option from the **Cutting Data** section. Next, click the **Select** button.
13. Click the **Geometry** tab. Next, select the edge chain of the model, as shown.

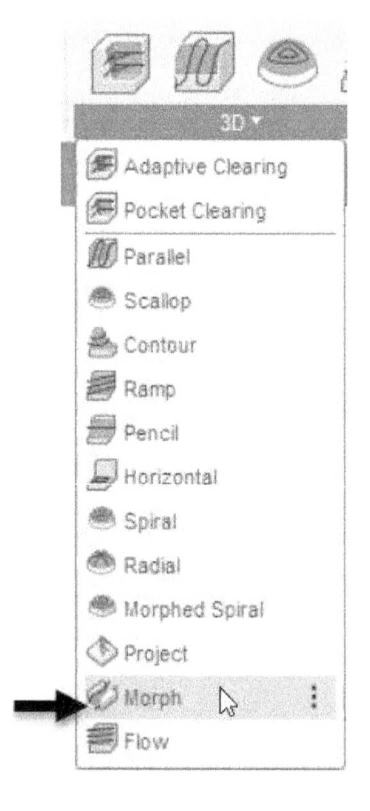

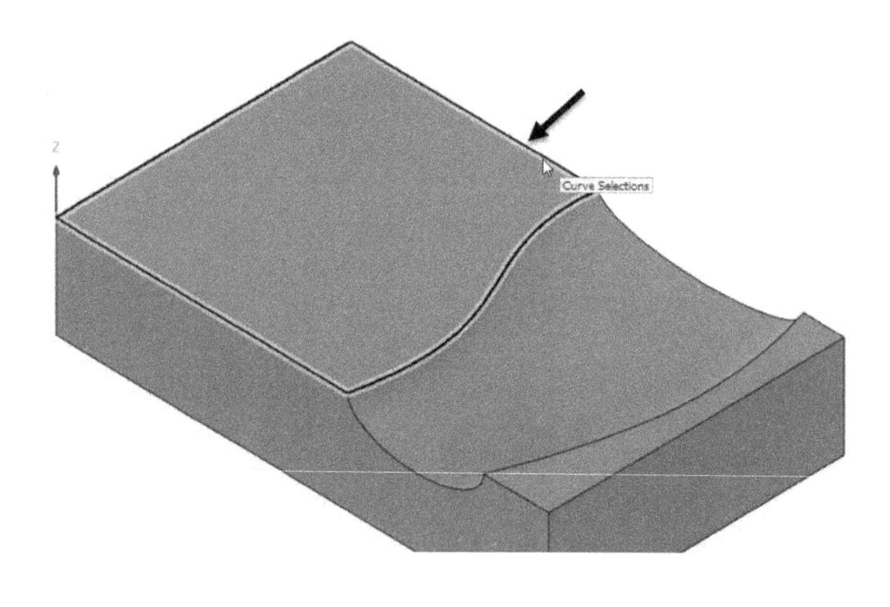

14. Click on the curved edge, as shown. Next, click the **Open Contour** icon.
15. Click the **Accept current contour** icon.

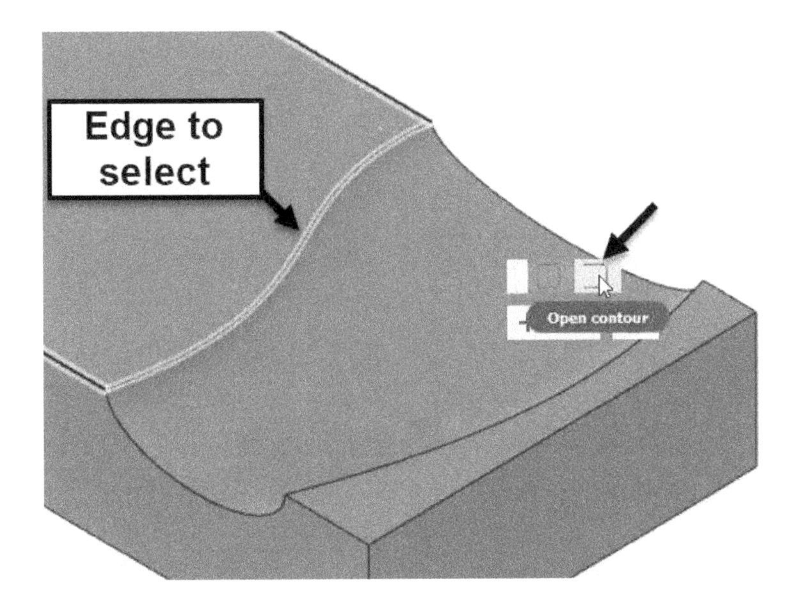

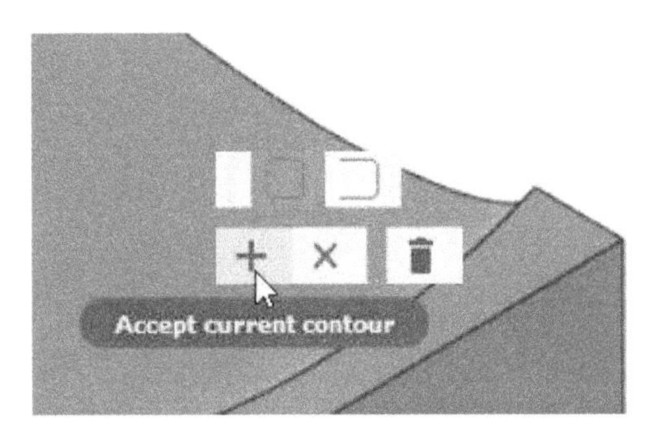

16. Select the edge chain, as shown. Next, click on the curved edge, as shown.

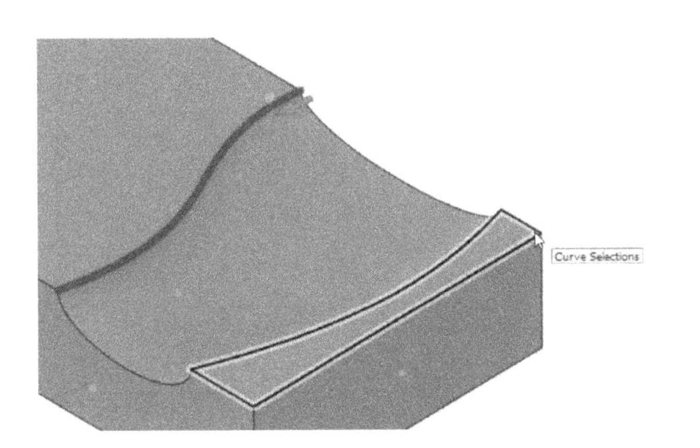

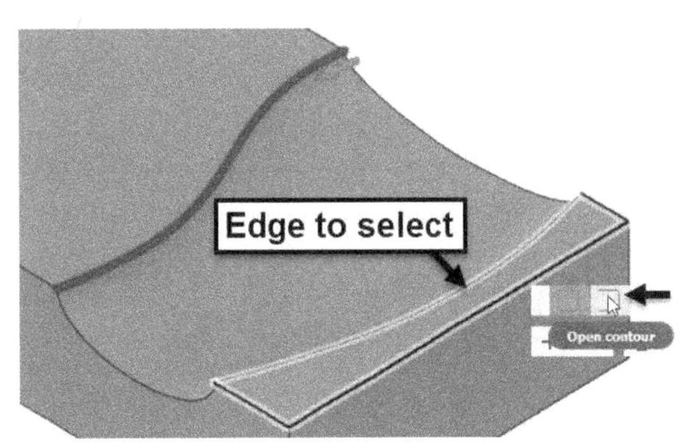

17. Click the **Open Contour** icon. Next, click the **Accept current contour** icon.
18. Click on the red arrow of the first curve to change the direction of the path.
19. Type **0.1** in the **Additional Offset** box.
20. Click the **Passes** tab on the dialog. Next, type **0.01** in the **Pass Extension** box.
21. Type **0.0125** in the **Stepover** box. Next, check the **Axial Offset Passes** option.
22. Type **0.0125** in the **Maximum Stepdown** box. Next, type **5** in the **Number of Stepdowns** box.
23. Click **OK**.
24. Simulate the operation and then close the **Simulate** dialog.

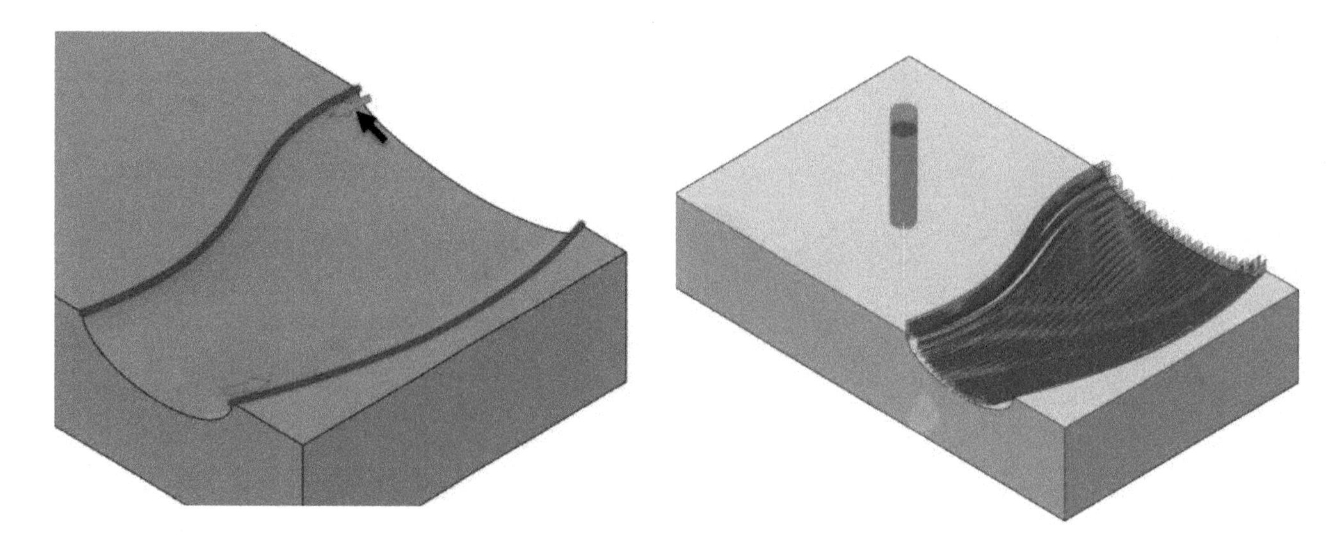

Creating the Morphed Spiral Operation

1. On the Toolbar, click **Milling > 3D > Morphed Spiral**.
2. On the **Morphed Spiral** dialog, select **Preset > Aluminum - Finishing**.
3. Click the **Geometry** tab. Next, check the **Rest Machining** option.
4. Under the **Rest Machining** section, select **Source > From previous operations(s)**.
5. Select **Adjustments > Machine cusps**.
6. Click the **Passes** tab on the dialog. Next, type **0.0125** in the **Stepover** box, and then click **OK**.
7. Simulate the operation and then close the **Simulate** dialog.

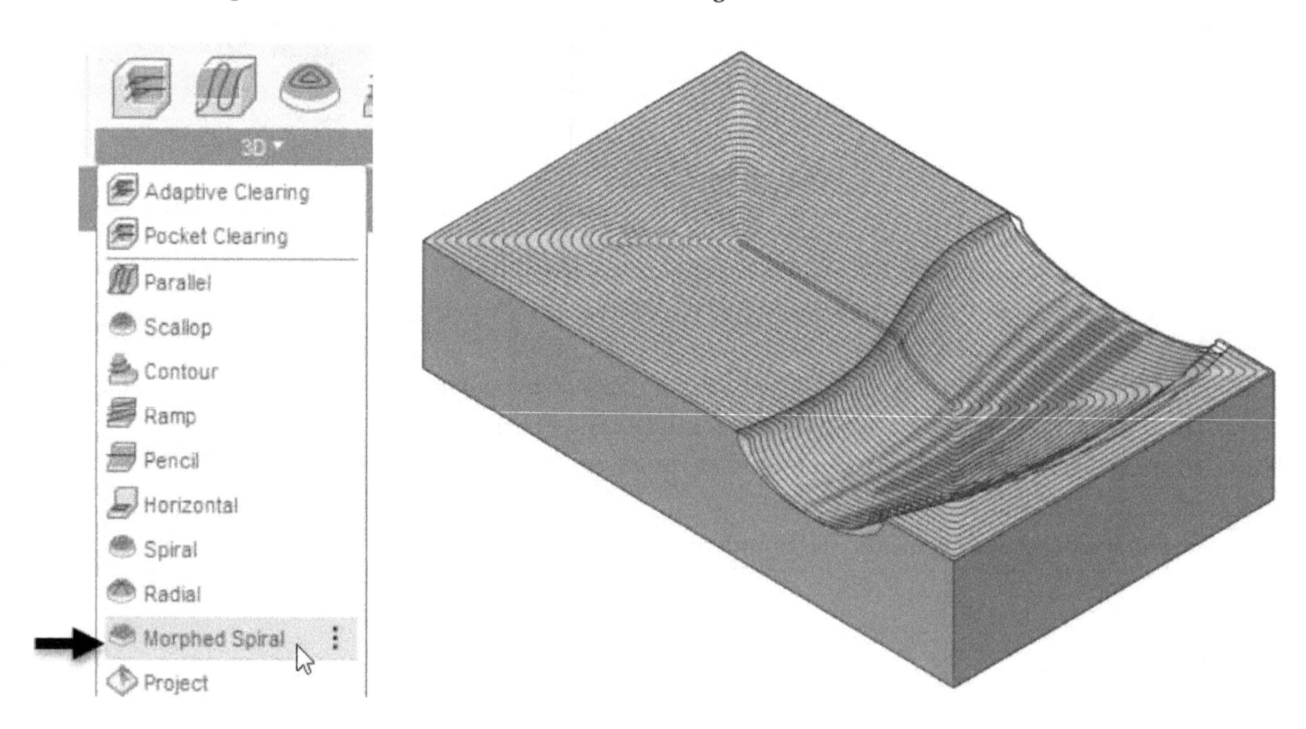

8. Save and close the file.

Tutorial 16

Creating the Ramp Operation

1. Download the **Tutorial 16** file, and then upload it to Fusion 360. Next, open the uploaded file.
2. On the Toolbar, click **Milling > 3D > Ramp.**
3. Click the **Select** button next to the **Tool** option.
4. Select the **Ø5/8" L0.015" (5/8" Bullnose Endmill)** tool from the tool list.
5. Select the **Aluminum - Finishing** option from the **Cutting Data** section. Next, click the **Select** button.
6. Click the **Geometry** tab. Next, select **Machining Boundary > Selection**.
7. Select the curved edge of the model, as shown.

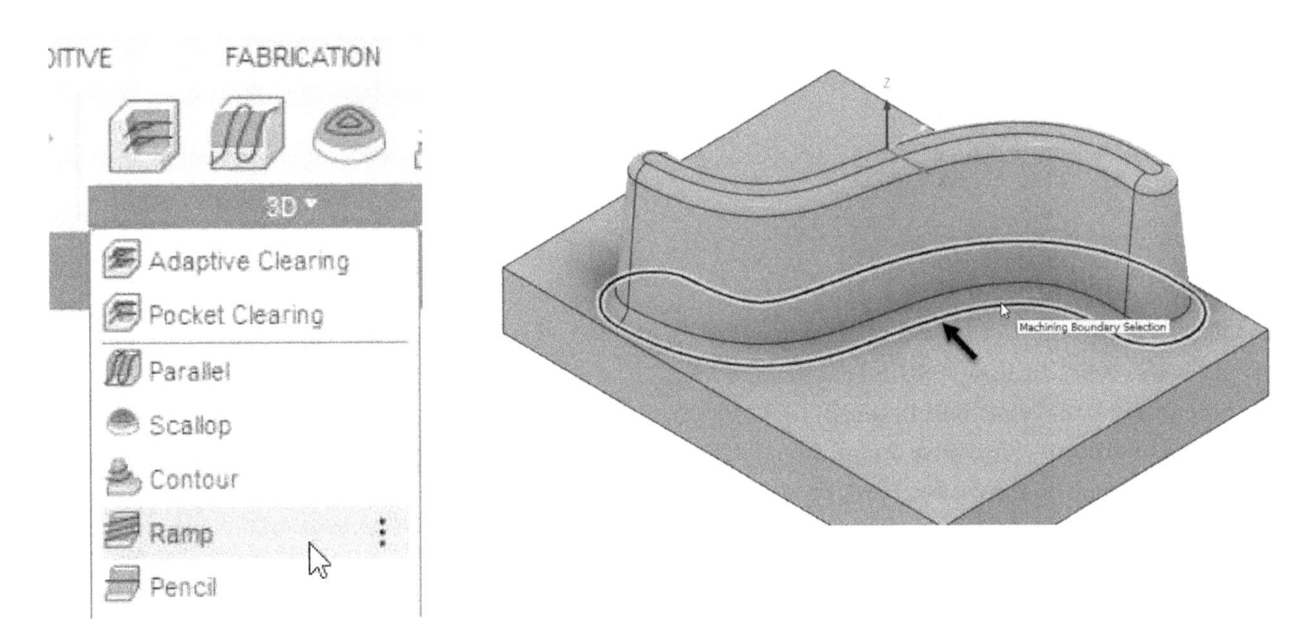

8. Type **0.5** in the **Additional Offset** box.
9. Click the **Heights** tab. In the **Top Height** section, select **From > Model Top**.
10. In the **Bottom Height** section, select **From > Selection**. Next, select the flat face of the model, as shown.
11. Click the **Passes** tab on the dialog. Next, type **0.01** in the **Maximum Stepdown** box.
12. Check the **Smoothing** option. Next, click **OK**.
13. Simulate the operation and then close the **Simulate** dialog.

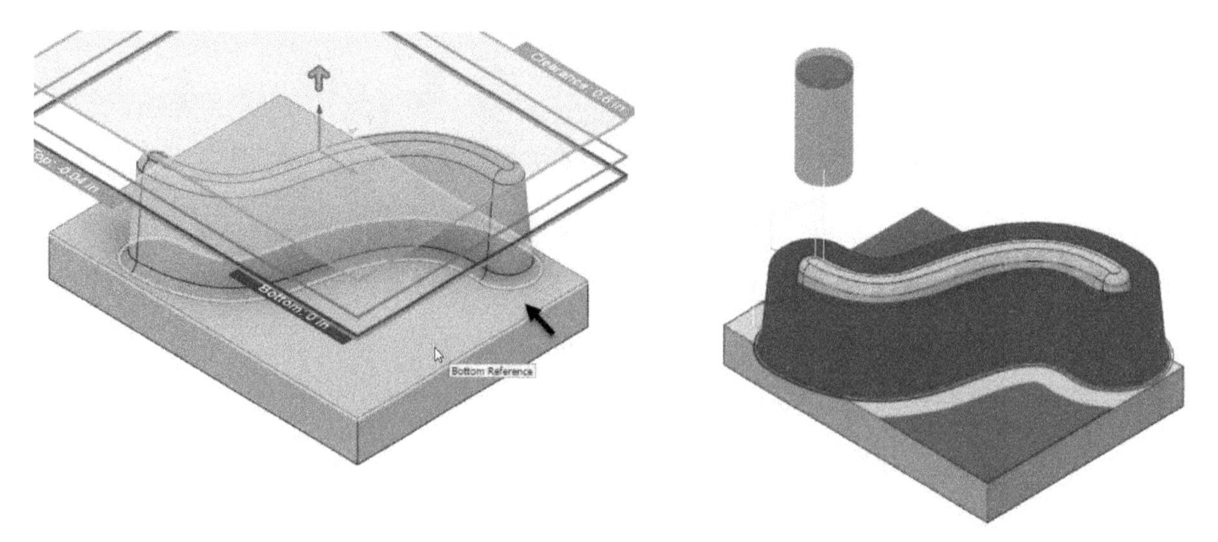

14. Save and close the file.

Tutorial 17

Creating the Pencil Operation

1. Download the **Tutorial 17** file, and then upload it to Fusion 360. Next, open the uploaded file.
2. On the Toolbar, click **Milling > 3D > Pencil.** Next, click the **Select** button next to the **Tool** option.
3. Select the **Ø1/4" L0.85" (1/4" Ball Endmill)** tool from the tool list.
4. Select the **Aluminum - Finishing** option from the **Cutting Data** section. Next, click the **Select** button.
5. Click the **Passes** tab on the dialog. Next, type **45** in the **Bitangency Angle** box.

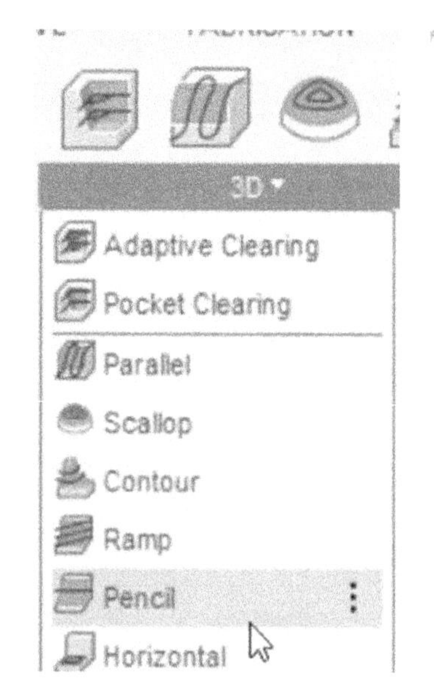

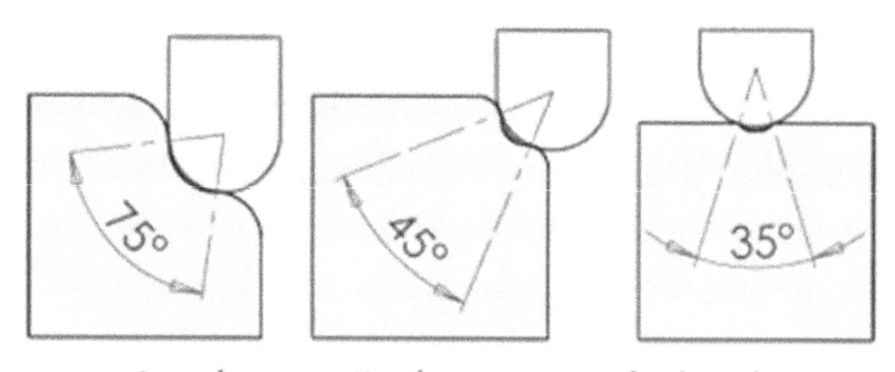

Cutter paths are created along all the contact areas where it would make two simultaneous contacts with the part, at more than this angular difference.

6. Type **2** in the **Number of Stepovers** box. Next, type **0.025** in the **Stepover** box, and then click **OK**.

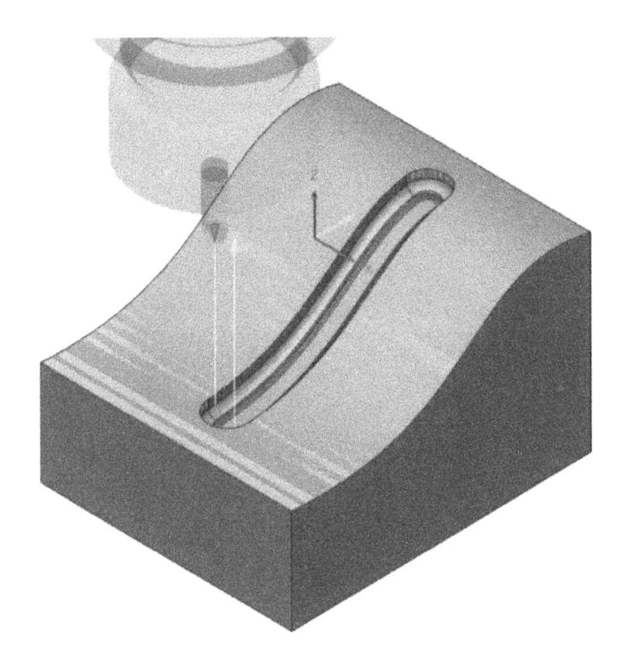

Tutorial 18

Creating the Project Operation

1. Download the **Tutorial 18** file, and then upload it to Fusion 360. Next, open the uploaded file.
2. On the Toolbar, click **Milling > 3D > Project.**

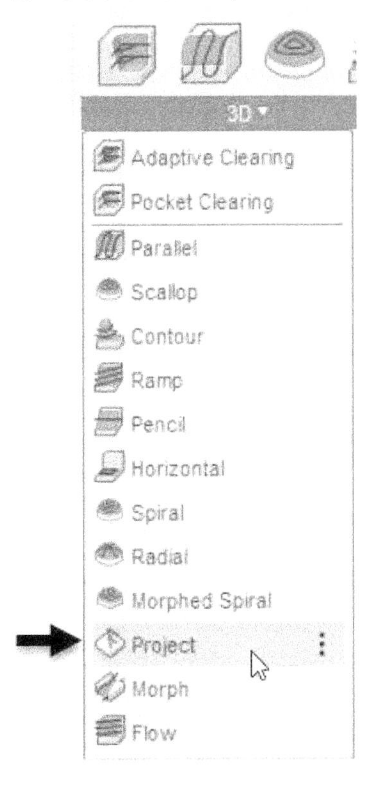

3. Click the **Select** button next to the **Tool** option.
4. Select the **Ø1/8" L0.6" (1/8" Ball Endmill)** tool from the tool list.
5. Select the **Aluminum - Finishing** option from the **Cutting Data** section. Next, click the **Select** button.

6. Click the **Geometry** tab. Next, select the elements of the sketch, as shown.
7. Click **OK**.

Tutorial 19

Creating the Flow Operation

The Flow operation generates a toolpath based on the Iso curves of a surface.

1. Download the **Tutorial 19** file, and then upload it to Fusion 360. Next, open the uploaded file.
2. On the Toolbar, click **Milling > 3D > Flow**. Next, click the **Select** button next to the **Tool** option.

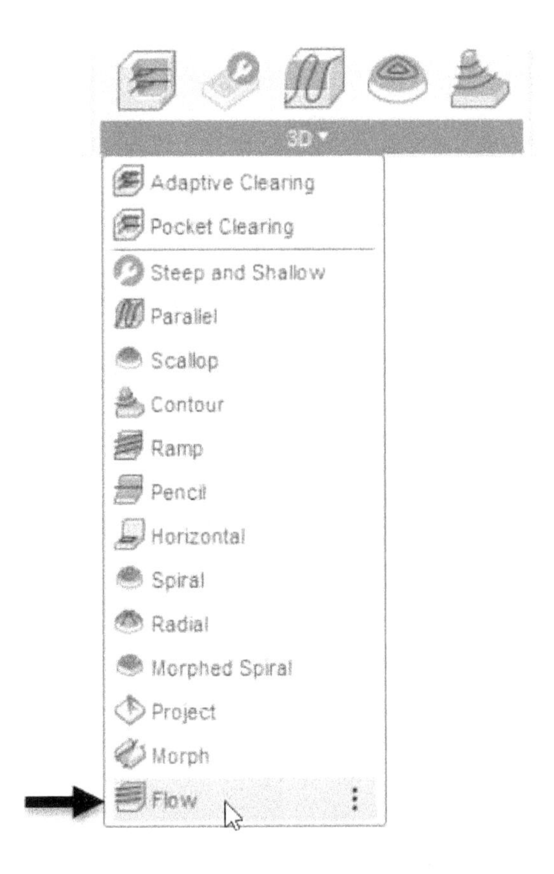

3. Select the **Ø1/8" L0.6" (1/8" Bullnose Endmill)** tool from the tool list.
4. Select the **Aluminum – Finishing** option from the **Cutting Data** section. Next, click the **Select** button.
5. Click the **Geometry** tab. Next, select the four faces of the model, as shown.
6. Click on the arrows displayed in the Y-direction; the flow direction is changed. Make sure that all the arrows point in the same direction.

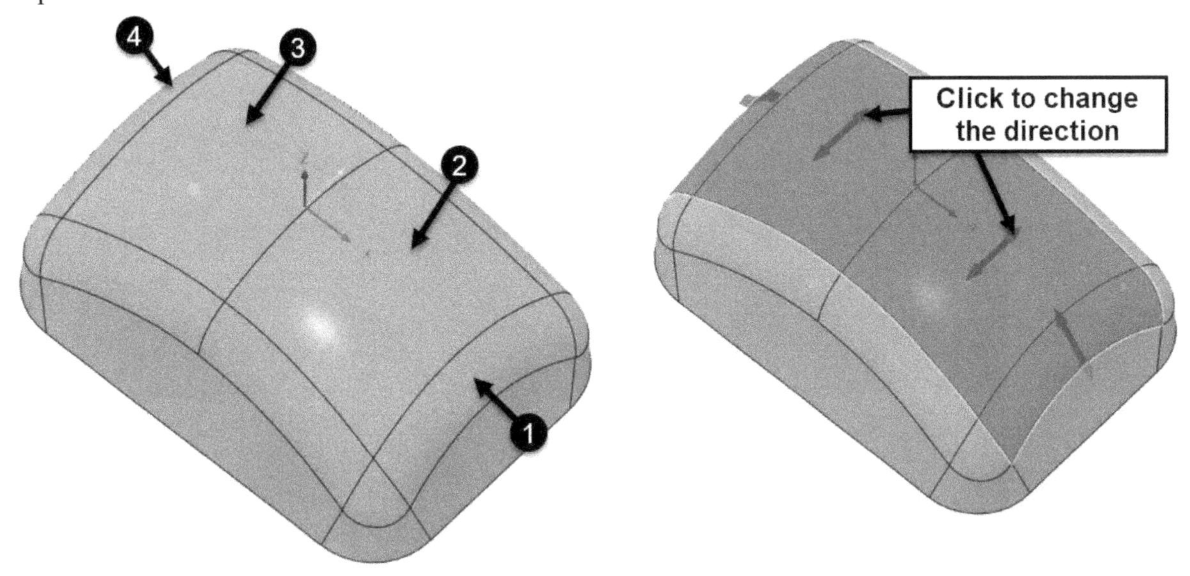

7. Click the **Passes** tab. Next, type **100** in the **Number of Stepovers** box.
8. Type **0.02** in the **Tangential Fragment Extension Distance** box.

9. Select **Direction > One Way**. Next, check the **Use Multi-axis** option.
10. Type **100** in the **Maximum Tilt** box.
11. Click the **Linking** tab. Next, select **Retraction Policy > Shortest path** and then click **OK**.
12. On the Toolbar, click **Milling > 3D > Flow**.
13. Click the **Geometry** tab. Next, select the faces of the model, as shown.

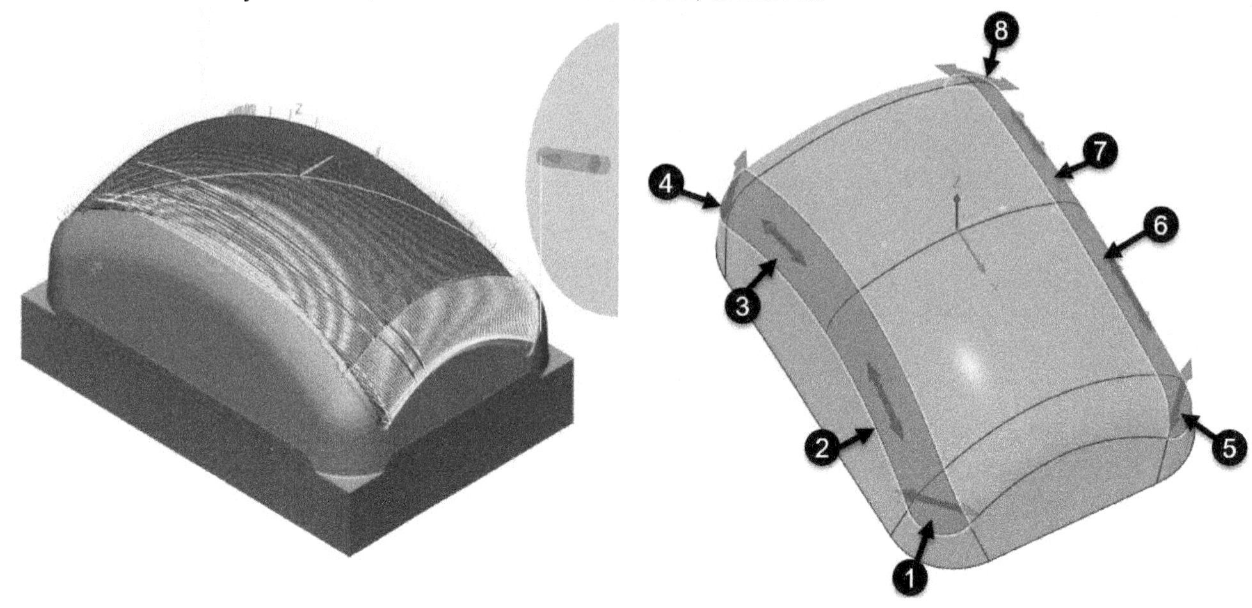

14. Click the **Passes** tab. Next, select **Isometric Direction > Along v**.
15. Type **30** in the **Number of Stepovers** box.
16. Type 0.02 in the **Tangential Fragment Extension Distance** box. Next, select **Direction > One Way**.
17. Check the **Use Multi-axis** option. Next, type **100** in the **Maximum Tilt** box.
18. Click the **Linking** tab. Next, select **Retraction Policy > Shortest path**, and then click **OK**.
19. Likewise, create the flow toolpath on the side faces, as shown.

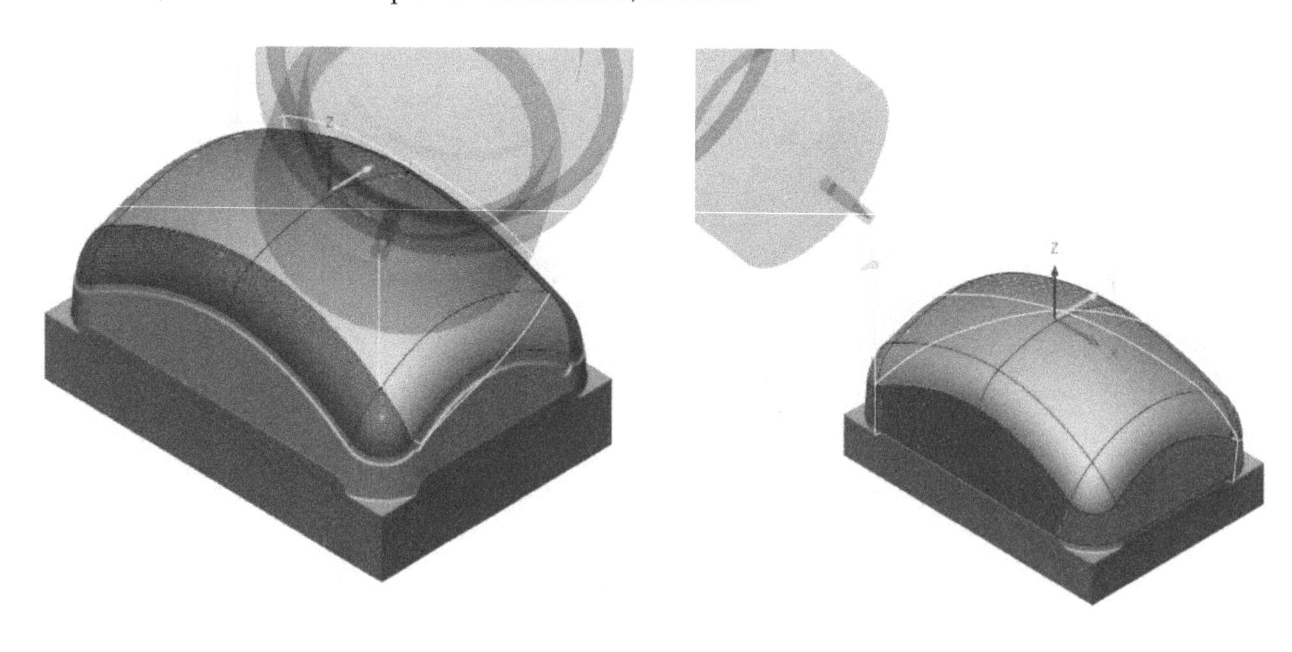

Simulating the Flow Operations

1. In the Browser, right-click on the Setup, and then select **Simulate**.
2. Click the **Go to next operation** button.

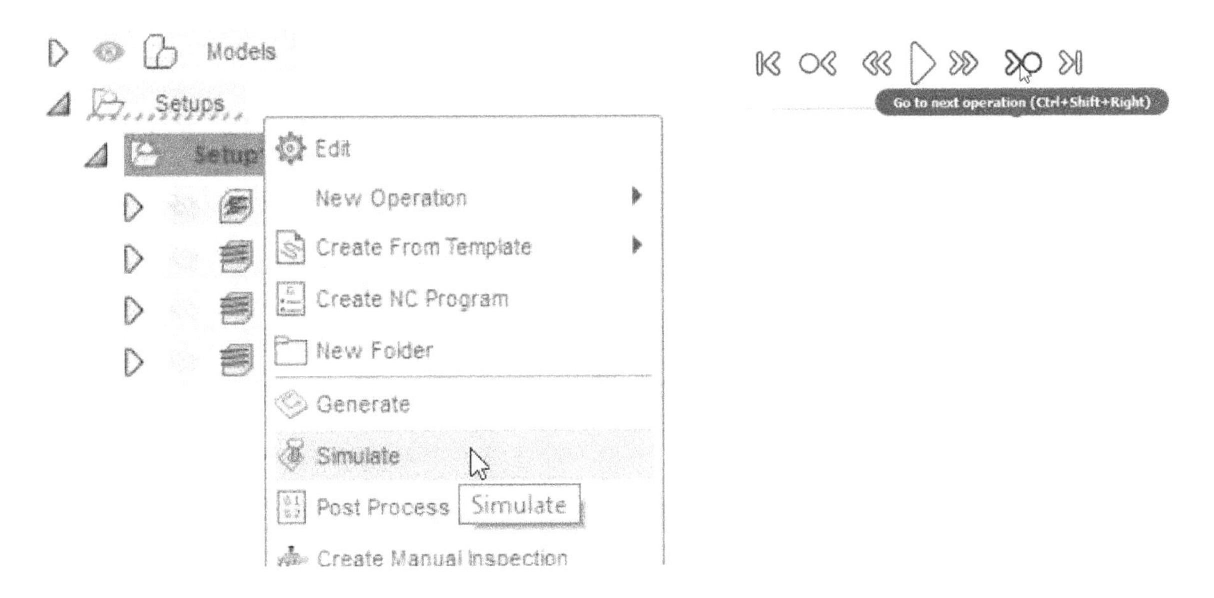

3. On the **Simulate** dialog, check the **Toolpath** option.
4. Check the **Show points** option under the **Toolpath** section. Notice the points on the toolpath. They are equally distributed along the toolpath. It will result in a smooth surface finish.
5. Click the **Start the simulation** button; the flow operations are simulated one-by-one.
6. Close the **Simulate** dialog after completing the simulation.

Tutorial 20

Creating the Swarf Operation

The **Swarf** operation is used to machine the inclined faces of the model using an endmill's side.

1. Download the **Tutorial 20** file, and then upload it to Fusion 360. Next, open the uploaded file.
2. On the Toolbar, click **Milling > Multi-axis > Swarf**. Next, click the **Select** button next to the **Tool** option.
3. Select the **Ø1/2" L1.1" (1/2" Flat Endmill)** tool from the tool list.
4. Select the **Aluminum – Finishing** option from the **Cutting Data** section. Next, click the **Select** button.
5. Click the **Geometry** tab. Next, click **Drive Mode > Contours**.

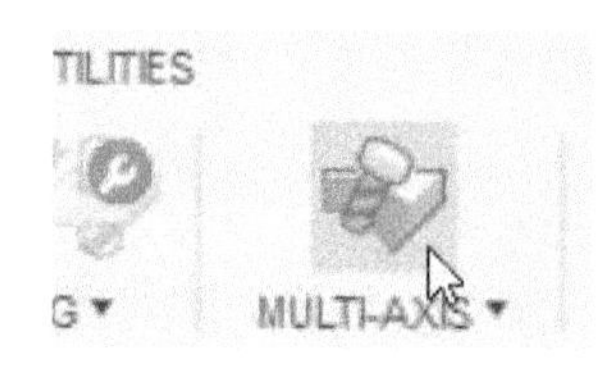

Drive Mode

Determines if the toolpath will be driven by the Faces of the model or user selected Contours.

With Faces selected, the actual surface will drive the Swarf Toolpath.

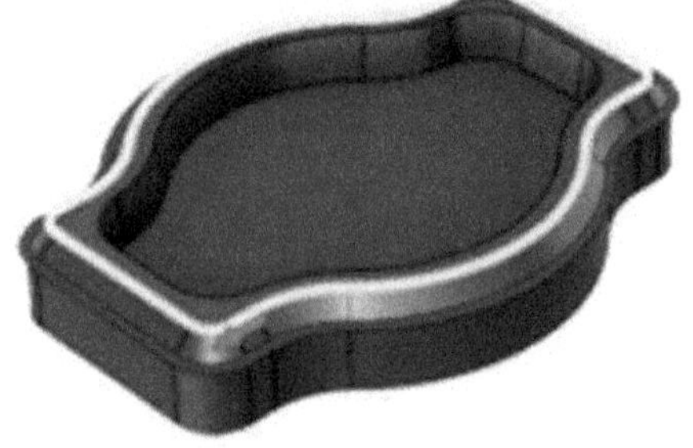

With Contours selected, the selected edges will drive the Swarf Toolpath.

6. Click **Selection Mode > Faces**.
7. Select the inclined face of the model, as shown; all the tangentially connected faces are selected. Note that the contours of the selected faces are used to generate the toolpath.
8. Click **OK** and notice that the toolpath is generated.
9. In the Browser, right-click on the **Swarf1** operation, and then select **Edit**.
10. Click the **Geometry** tab. Next, click **Selection Mode > Contour pairs**.
11. Select the lower contour and the top contour, as shown.

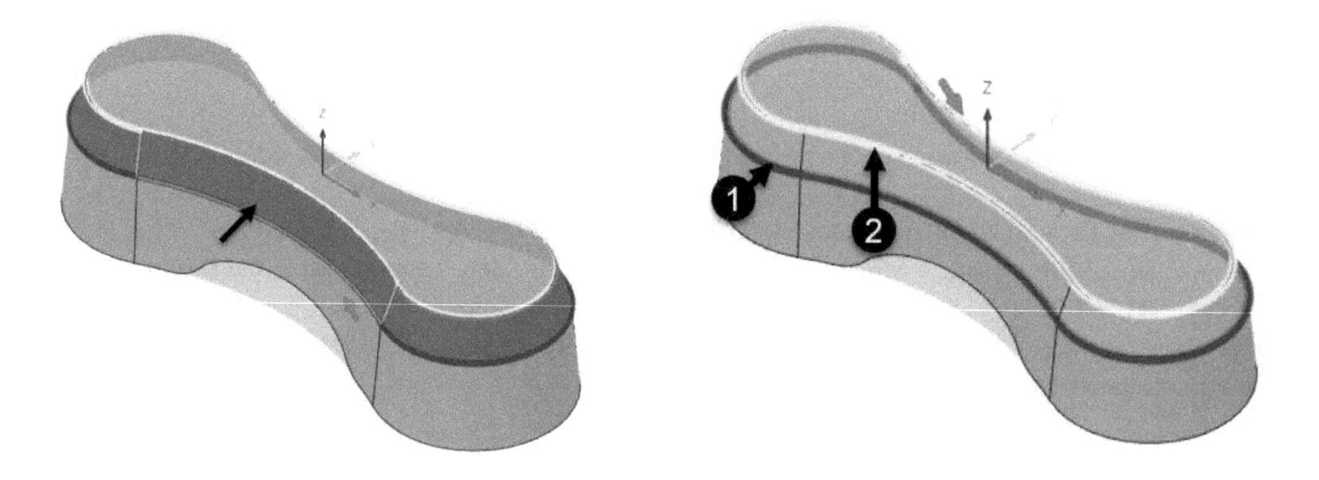

The bottom of the tool will touch the first selected contour. The second selected contour defines the inclination angle of the tool.

12. Click the **Linking** tab.
13. In the **Positions** section, click the selection button next to the **Entry Positions** option.
14. Select the vertex point, as shown.

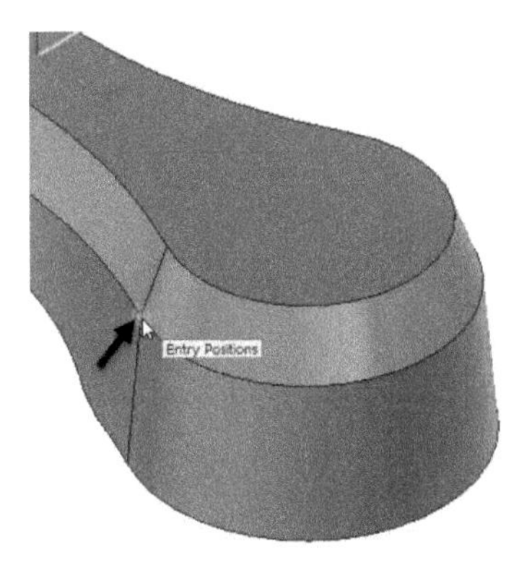

15. Click **OK**; the entry and exit positions of the toolpath are defined, as shown.

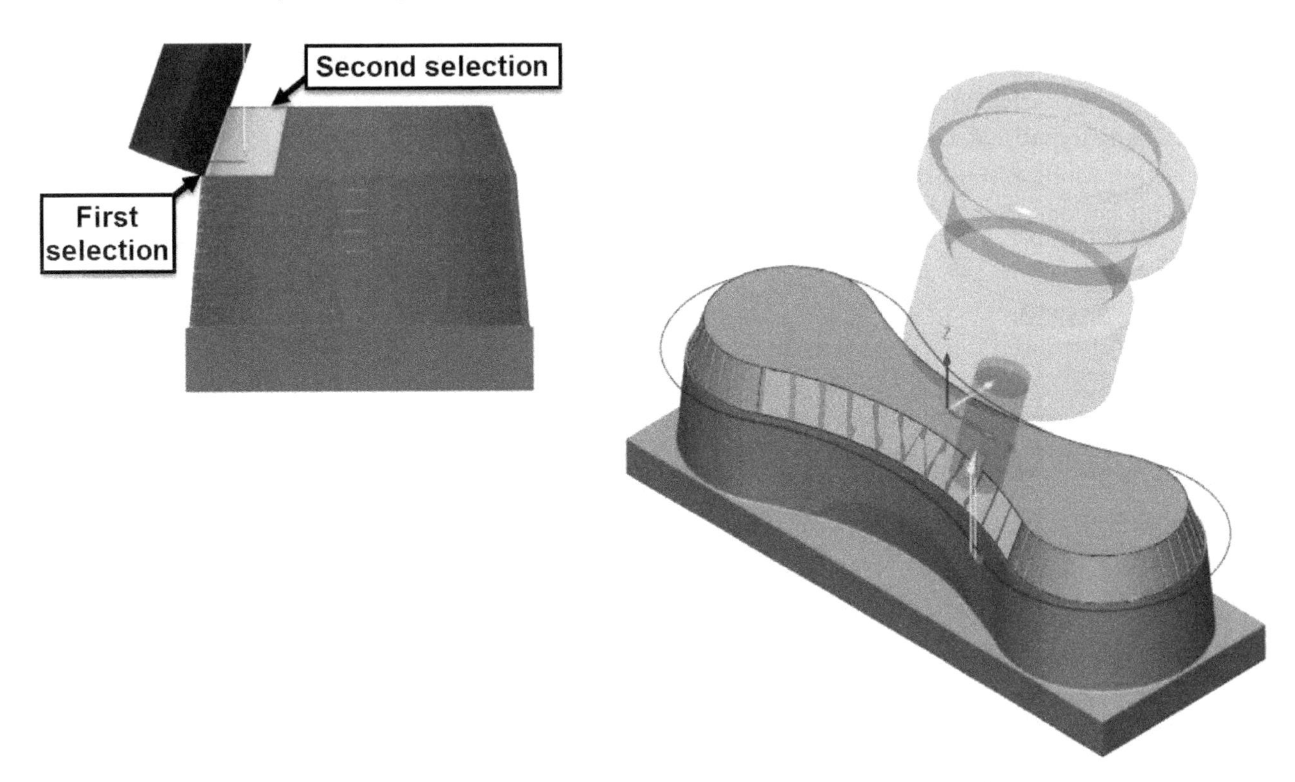

16. On the Toolbar, click **Milling > Multi-axis > Swarf**. Next, click the **Geometry** tab.
17. Click **Drive Mode > Surfaces**. Next, click **Selection Mode > Manual**.
18. Select the two faces, as shown. Next, click **OK** and notice that partial swarf is created.

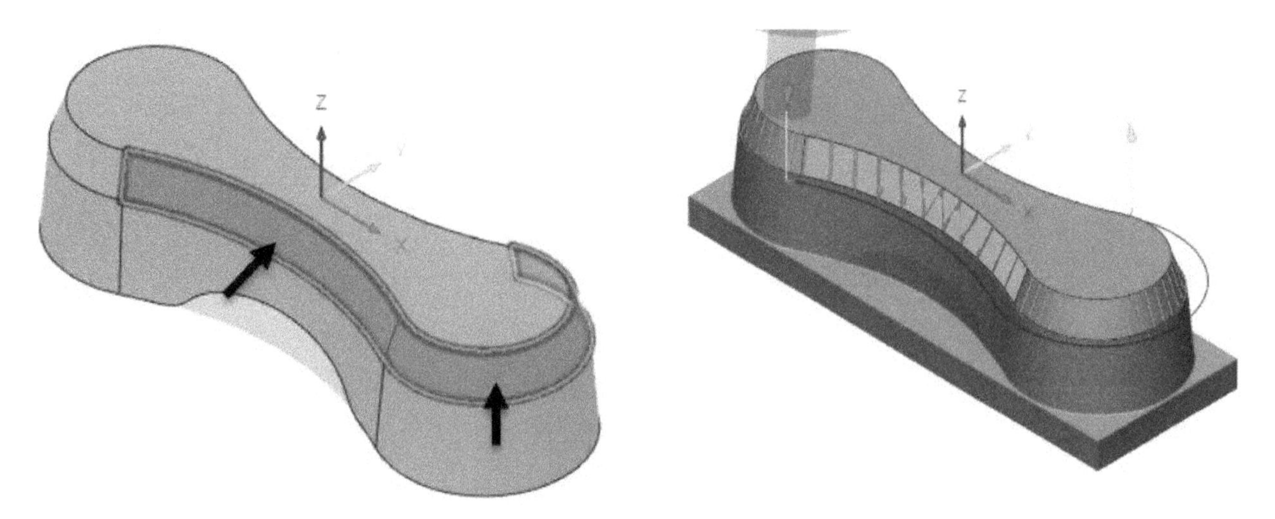

19. In the Browser, right-click on the **Swarf2** operation, and then select **Edit**.
20. Click the **Geometry** tab. Next, click **Drive Mode > Contour**.
21. Click **Selection Mode > Manual**. Next, select the edges in the sequence, as shown.
22. Click **OK**; the bottom edge of the tool will touch the first selected edge.

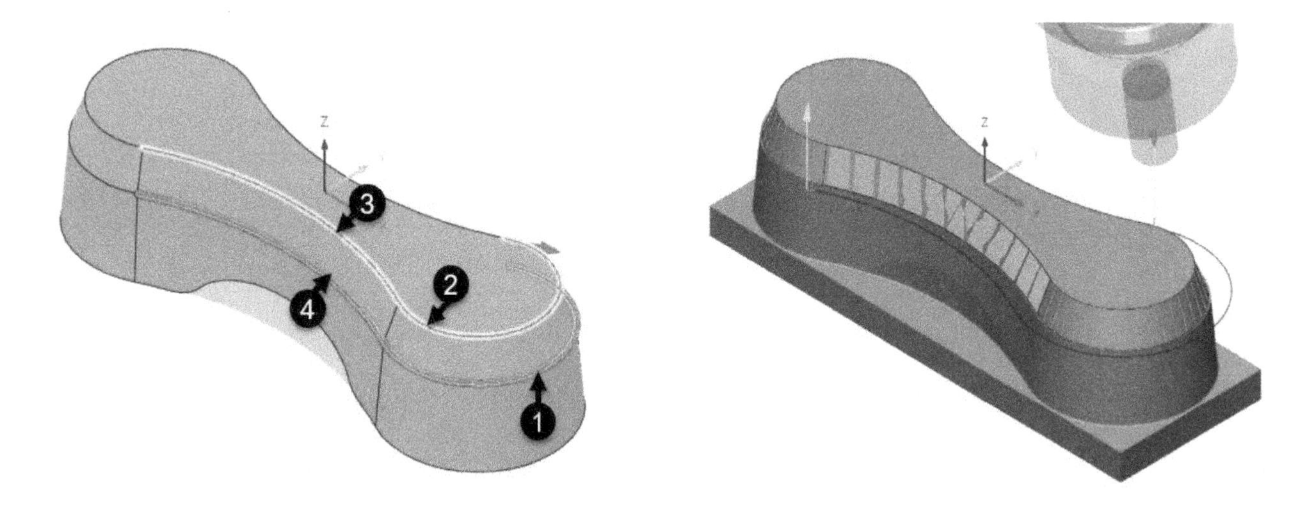

23. On the Toolbar, click **Milling > Multi-axis > Swarf**. Next, click the **Select** button next to the **Tool** option.
24. Right-click on the **Ø1/2" L1.1" (1/2" Flat Endmill)** tool. Next, select the **Duplicate tool** option.
25. Select the **Edit tool** option.

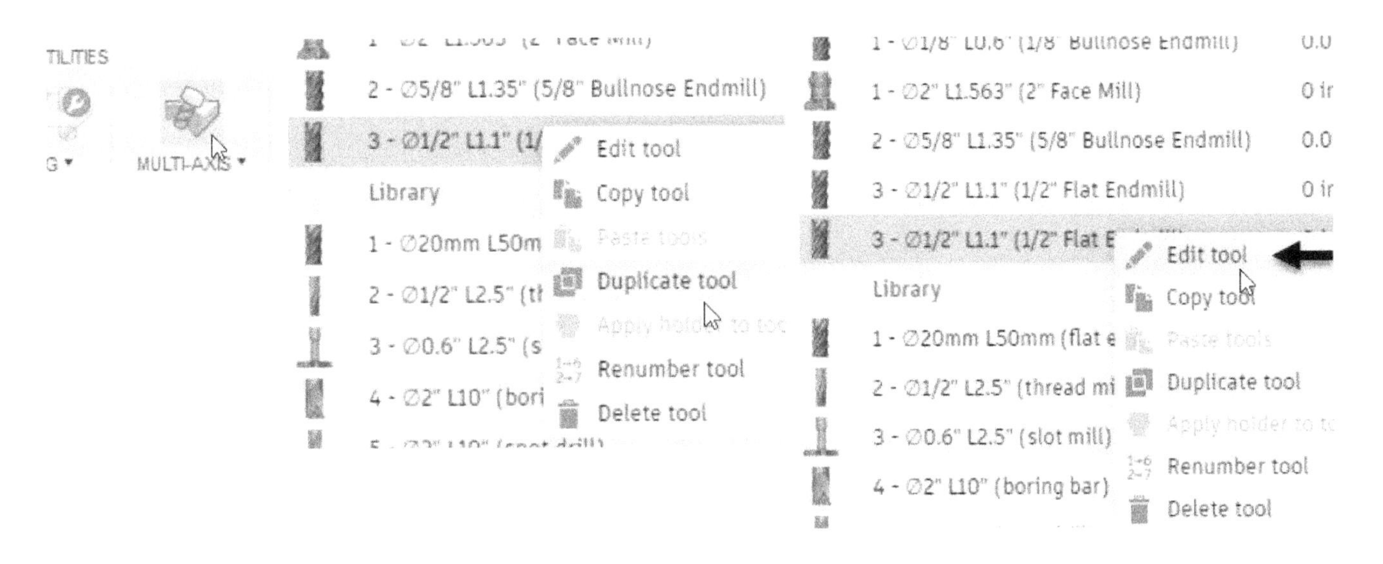

26. Click the **Cutter** tab.
27. Enter the following values in the **Geometry** section.

***Diameter:* 0.5**
***Shaft diameter:* 0.5**
***Overall length:* 5**
***Length below holder:* 3**
***Shoulder length:* 2**
***Flute length:* 2**

28. Click the **Accept** button. Next, click the **Select** button.
29. Select the **Aluminum – Finishing** option from the **Cutting Data** section. Next, click the **Select** button.
30. Click the **Geometry** tab. Next, click **Drive Mode > Contours**.
31. Click **Selection Mode > Faces**. Next, select the lower inclined face of the model, as shown.
32. Click the **Passes** tab. Next, click **Cutting Mode > From Top**.
33. Type **0.1** in the **Maximum Stepdown** box. Next, click **OK**.

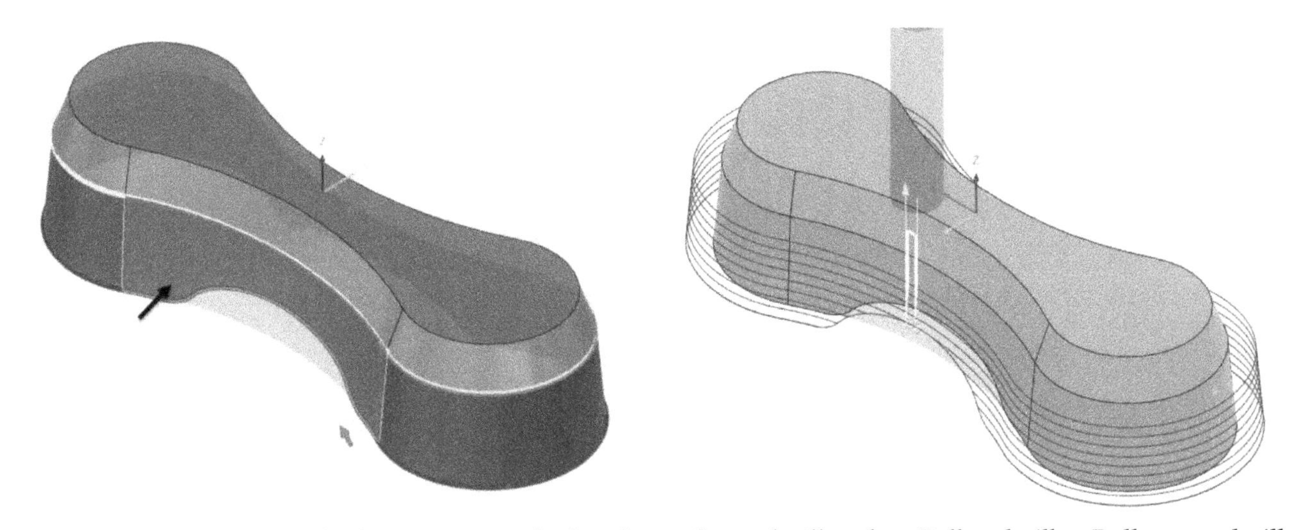

Multiple passes can be used when using a tool other than a flat endmill such as Ball endmill or Bullnose endmill.

Different Cutting Modes for multiple passes are explained in the following figures.

From Bottom

Trim From Bottom

Trim From Top

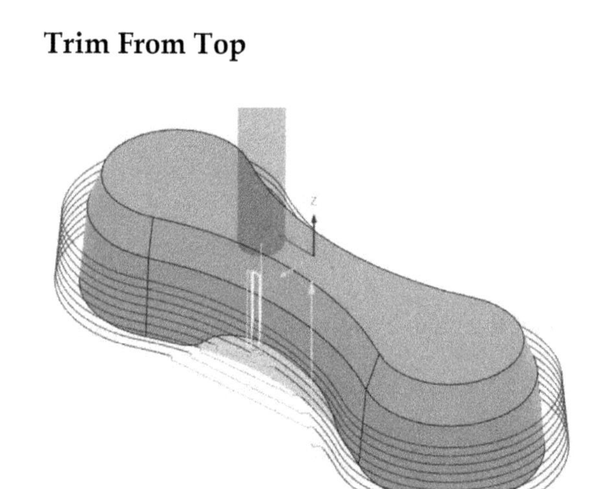

Spiral

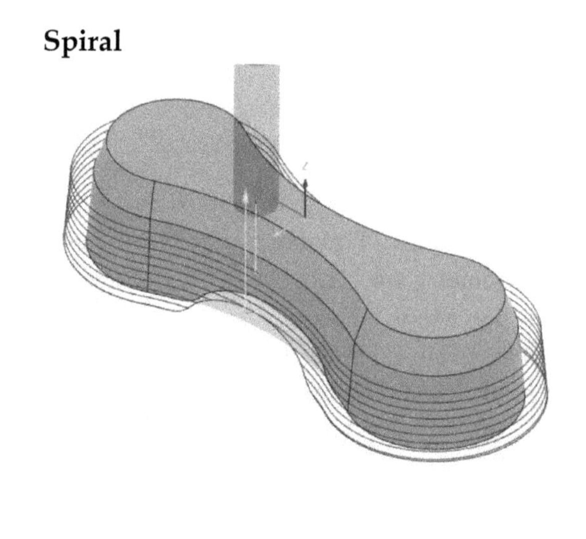

Morph

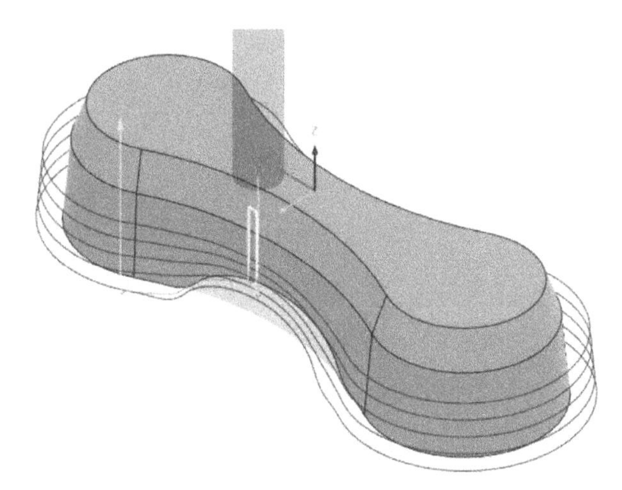

34. Simulate the operations.
35. Save and close the file.

Tutorial 21

Creating the Wrap Toolpath

The **Wrap Toolpath** option is used to machine the pockets that are wrapped around a cylindrical face.

1. Download the **Tutorial 21** file and then upload it to Fusion 360. Next, open the uploaded file.
2. On the Toolbar, click **Milling > Setup > Setup**. On the **Setup** dialog, select **Operation Type > Milling**.
3. On the **Setup** dialog, click the **Stock** tab. Next, click **Mode > Relative size tube**.
4. Select the cylindrical face of the model to define the axis of the stock.

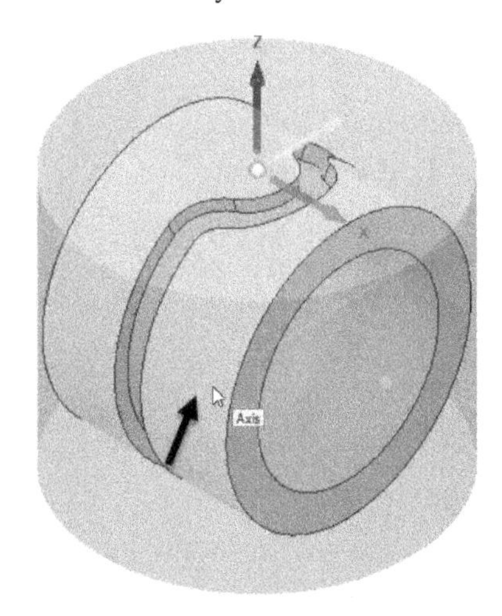

5. Type **0** in the **Radial Stock Offset** box.
6. Click the **Setup** tab. On the **Setup** tab, select **Orientation > Select Z axis/plane & X axis**.
7. Select the vertical axis of the coordinate system to define the Z axis.
8. Select the cylindrical face to define the X axis.
9. Check the **Flip X Axis** option if the X-axis is pointing towards the left.
10. Select the **Box point** button next to the **Stock point** option. Next, select the stock point, as shown.
11. Click **OK** on the **Setup** dialog.
12. On the Toolbar, click **Milling > 2D > 2D Adaptive Clearing**.

13. On the **2D Adaptive** dialog, click the **Select** button next to the **Tool** option.
14. On the **Select Tool** window, select the **Sample Tools - Inch** folder from the **Fusion 360 Library**.
15. Select the **Ø3/16" L0.725" (3/16" Flat Endmill)** tool from the tool list.
16. Click the **Select** button located at the bottom-right corner.
17. Click the **Geometry** tab on the **2D Adaptive** dialog.
18. Check the **Wrap Toolpath** option.
19. Select the cylindrical face of the model.

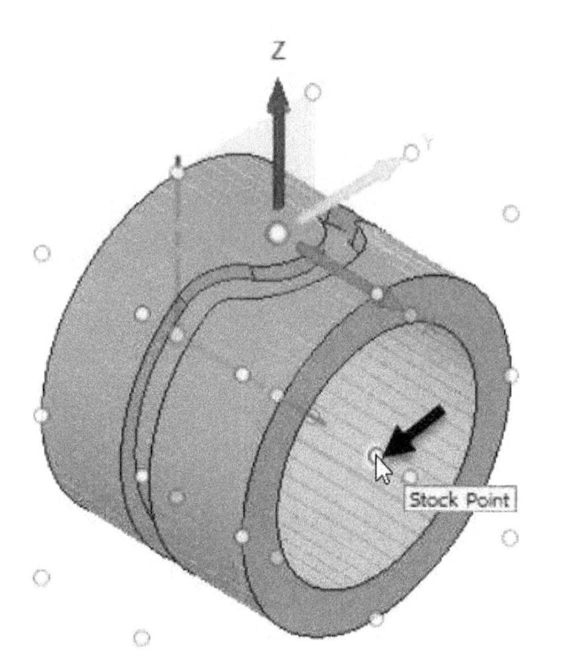

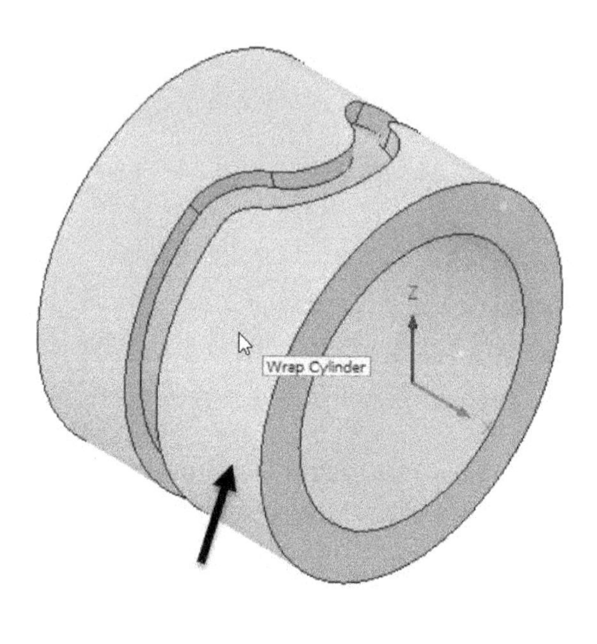

20. Click the **Pocket Selections** button. Next, select the contour of the pockets, as shown.
21. Click the **Passes** tab. Next, right-click in the **Optimal Load** box, and then select **Edit Expression**.
22. Type **tool_diameter * 0.2** in the **Edit Expression** dialog and click **OK**.
23. In the **Stock to Leave** section, type **0** in the **Radial Stock to Leave** and **Axial Stock to Leave** boxes.
24. Click **OK** to generate the toolpath.

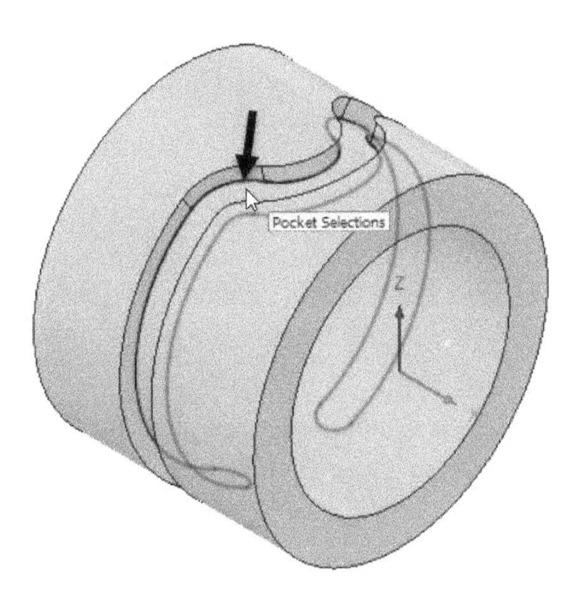

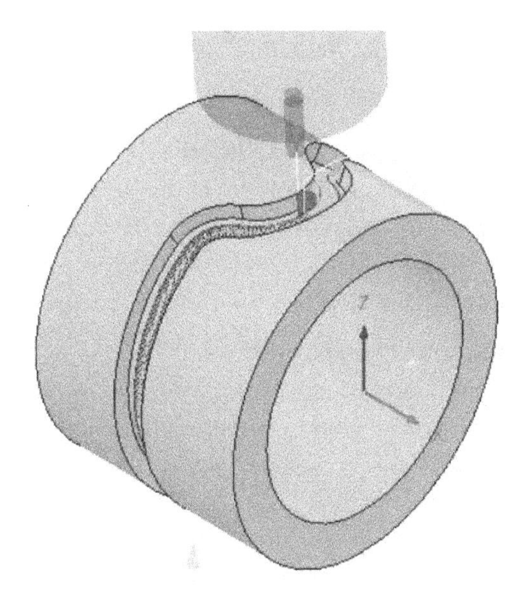

25. Simulate the toolpath.
26. Save and close the file.

Tutorial 22

Using the Tool Orientation Option

The **Tool Orientation** option helps you to add another Work Coordinate System to the setup.

1. Download the **Tutorial 22** file and upload it to Fusion 360. Next, open the uploaded file.
2. On the Toolbar, click **Milling > Setup > Setup**. On the **Setup** dialog, select **Operation Type > Milling**.
3. On the **Setup** dialog, click the **Stock** tab. Next, click **Mode > Relative size cylinder**.
4. Select the cylindrical face of the model to define the axis of the stock.

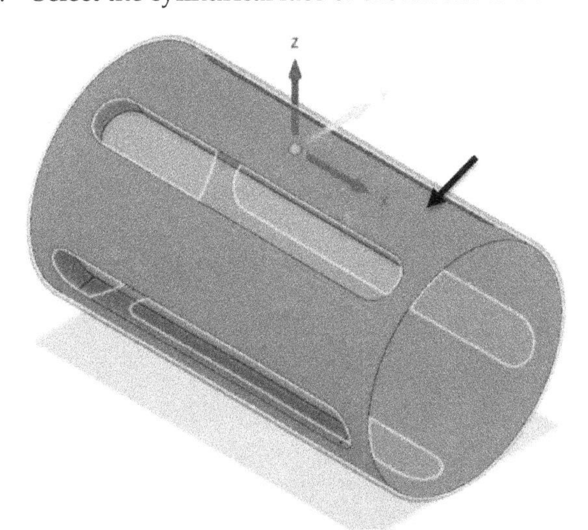

5. Type **0** in the **Radial Stock Offset** box. Next, click the **Setup** tab.
6. On the **Setup** tab, select **Orientation > Select Z axis/plane & X axis**.
7. Select the vertical axis of the coordinate system to define the Z axis.
8. Select the cylindrical face to define the X axis.
9. Select the **Box point** button next to the **Stock point** option. Next, select the stock point, as shown.

10. Click **OK** on the **Setup** dialog.
11. On the Toolbar, click **Milling > 2D > 2D Adaptive Clearing**.
12. On the **2D Adaptive** dialog, click the **Select** button next to the **Tool** option.
13. On the **Select Tool** window, select the **Sample Tools - Inch** folder from the **Fusion 360 Library**.
14. Select the **Ø1/8" L0.6" (1/8" Flat Endmill)** tool from the tool list.
15. Click the **Select** button located at the bottom-right corner.
16. Click the **Geometry** tab on the **2D Adaptive** dialog.
17. Check the **Tool Orientation** option.
18. Select **Tool Orientation > Select Z-axis/plane & X-axis.**
19. Select the flat face of the slot feature; the Z-axis is defined perpendicular to the selected face.
20. Click on the Z-axis to reverse the direction of the Z-axis if the Z-axis is pointing downwards.

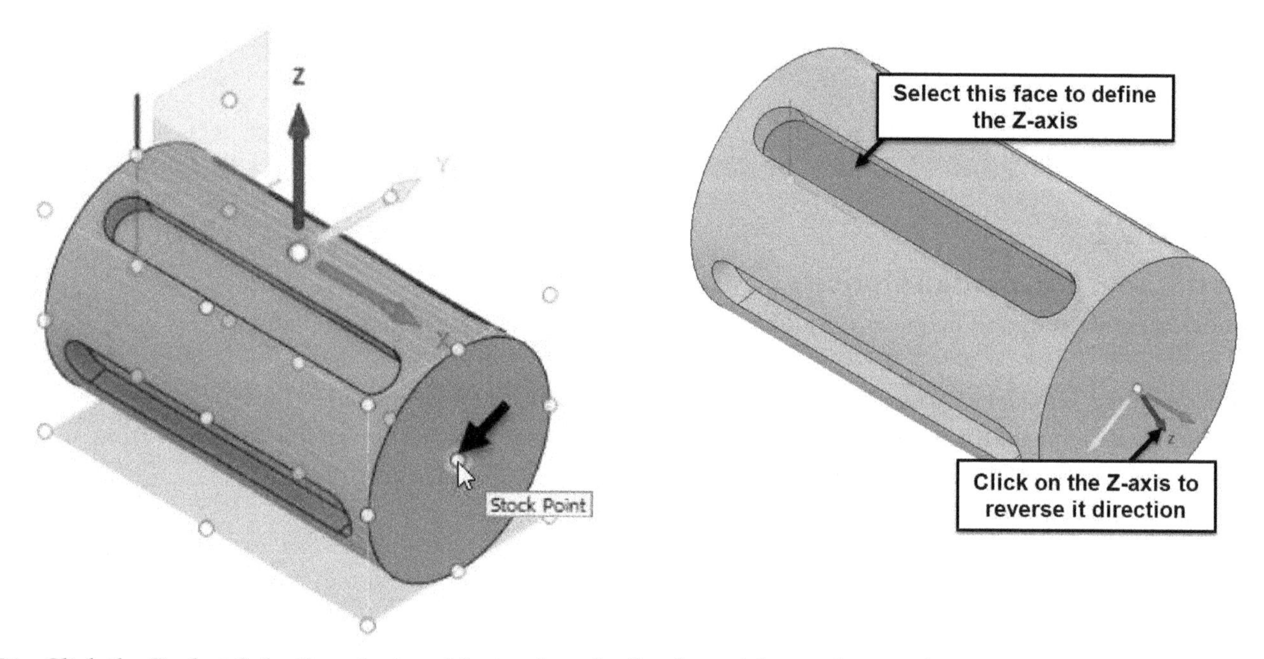

21. Click the **Pocket Selections** button. Next, select the flat face of the pocket, as shown.
22. Click the **Passes** tab. Next, right-click in the **Optimal Load** box, and then select **Edit Expression.**
23. Type **tool_diameter * 0.2** in the **Edit Expression** dialog and click **OK.**
24. In the **Stock to Leave** section, type **0** in the **Radial Stock to Leave** and **Axial Stock to Leave** boxes.
25. Click **OK** to generate the toolpath.

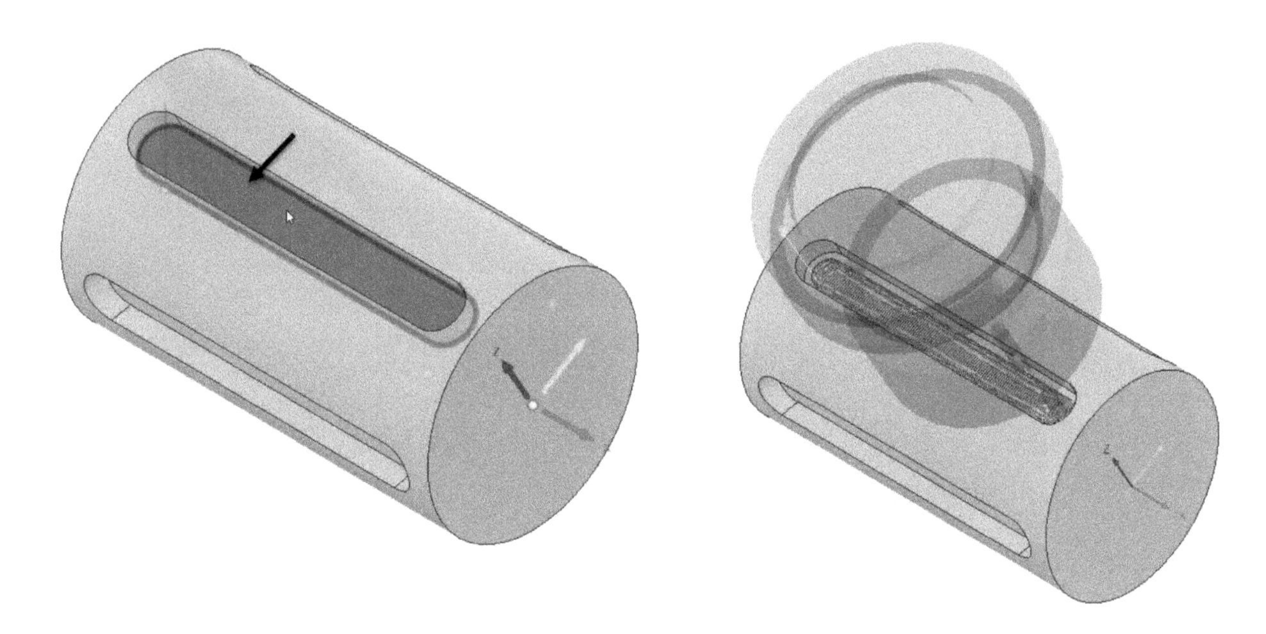

Creating the Circular Pattern of the toolpaths

1. Select the **2D Adaptive** operation from the Browser. Next, right-click and select **Add to New Pattern**.
2. On the **Pattern** dialog, select **Pattern Type > Circular pattern**.
3. Select the cylindrical face of the model to define the axis of the circular pattern.

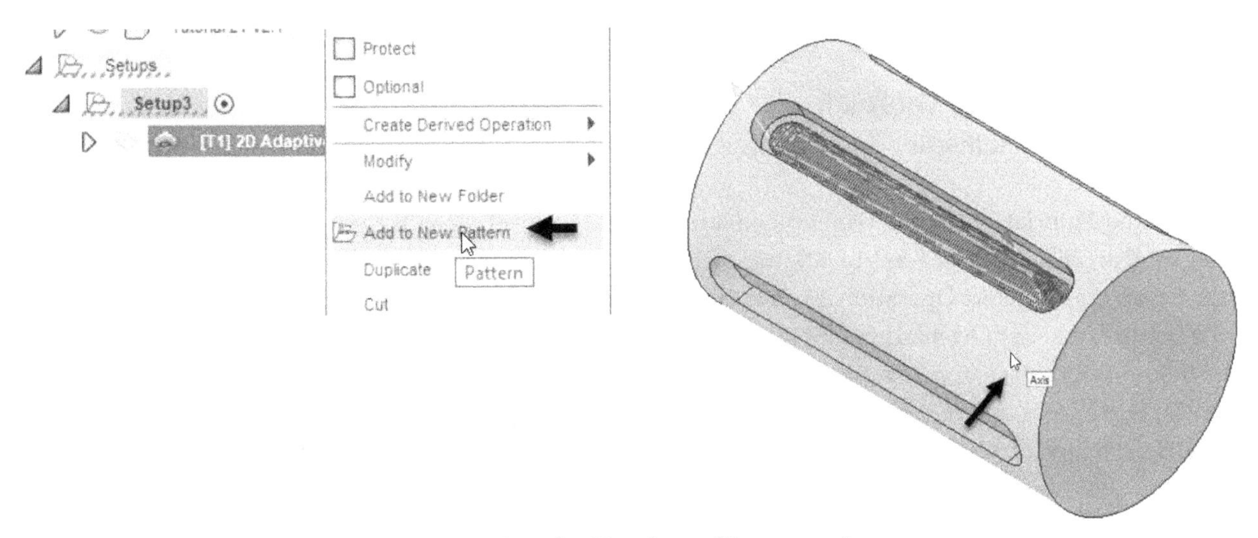

4. Type **360** in the **Angle** box. Next, type **5** in the **Number of Instances** box.
5. Check the **Equal spacing** option. Next, click **OK**.
6. Simulate the pattern of toolpaths.

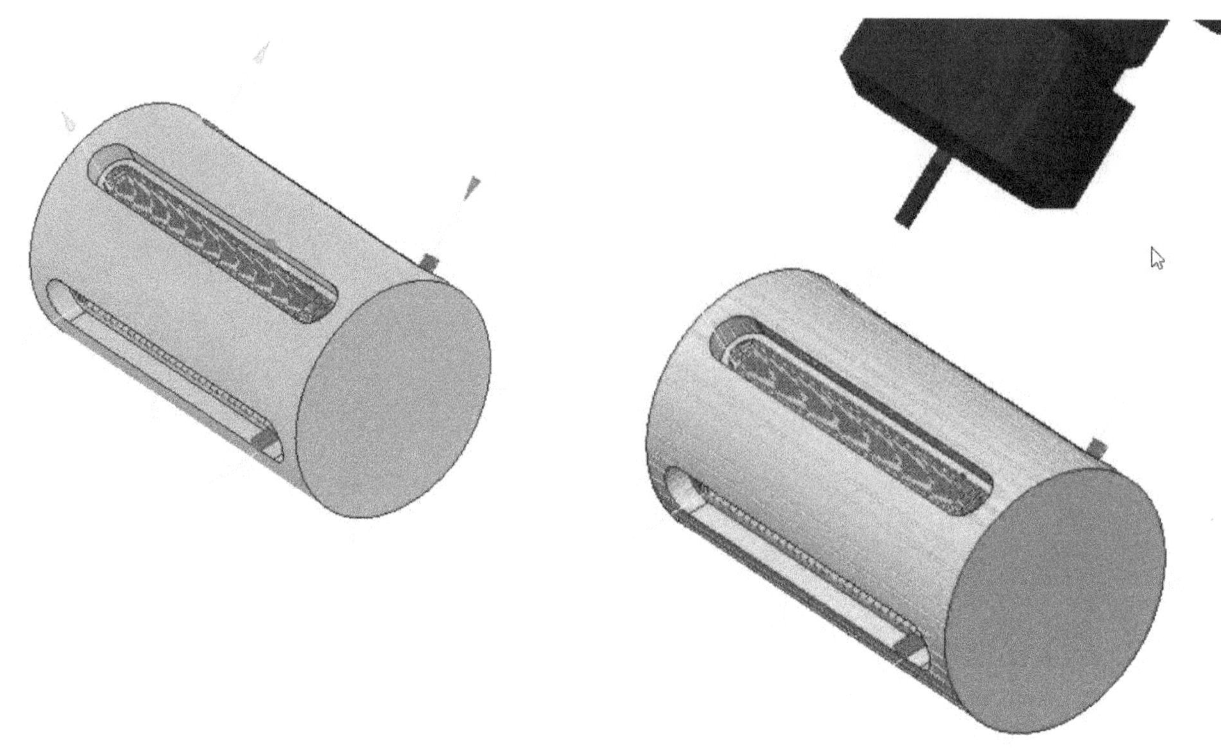

7. Save and close the file.

Tutorial 23

Creating the 4-axis Rotary Toolpath

The **Rotary** operation is a multi-axis operation that is used to machine a part around and along a rotary axis.

1. Download the **Tutorial 22** file and then upload it to Fusion 360. Next, open the uploaded file.
2. On the Toolbar, click **Milling > Setup > Setup**.
3. On the **Setup** dialog, select **Operation Type > Turning or mill/turn**.
4. On the **Setup tab**, select **Orientation > Select Z Axis/plane & X axis**.
5. Select the cylindrical face to define the Z axis.
6. Check the **Flip Z Axis** option if the Z-axis is pointing towards the left.
7. Select **Origin > Stock front**.

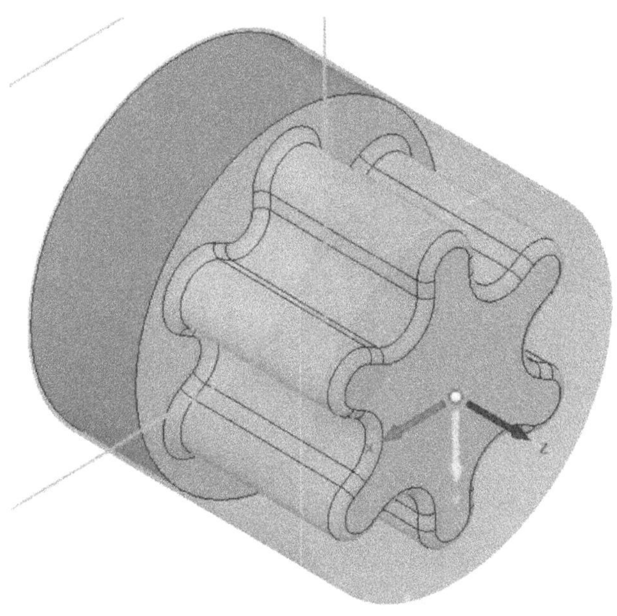

8. Check the **Spun Profile** option. On the **Setup** dialog, click the **Stock** tab.
9. Click **Mode > Relative size cylinder**.
10. Type **0.04** in the **Radial Stock Offset and Frontside Stock Offset** boxes. Next, click **OK**.
11. On the Toolbar, click **Turning > Turning > Turning Face**.
12. On the **Face** dialog, click the **Select** button next to the **Tool** option.
13. On the **Select Tool** window, select the **Turning – Sample Tools** folder from the **Fusion 360 Library**.
14. Select the **CNMT (CNMT Right Hand)** tool from the tool list.
15. Click the **Select** button located at the bottom-right corner. Next, click **OK** on the **Face** dialog.

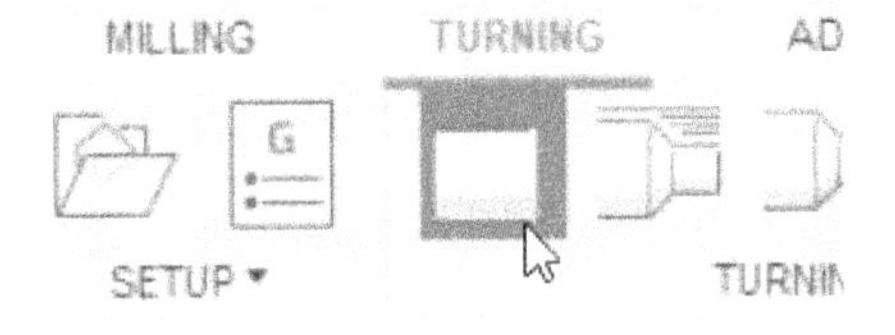

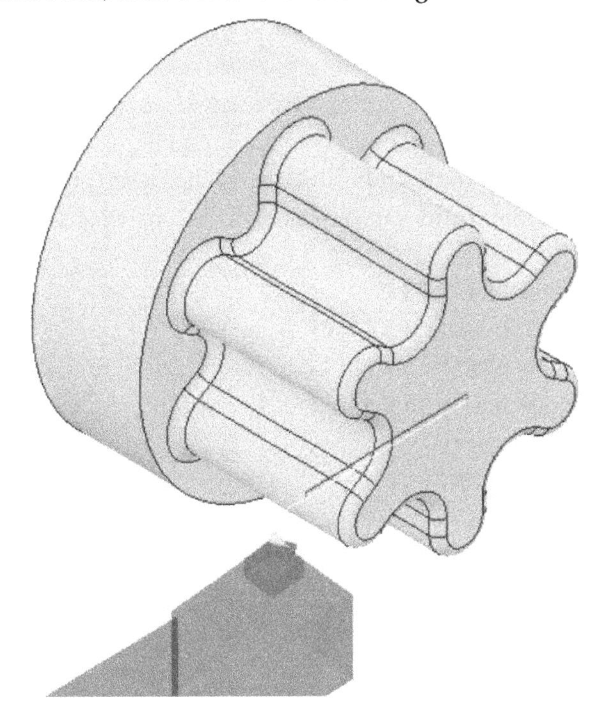

16. On the Toolbar, click **Turning > Turning > Turning Profile Roughing**.
17. Leave the default settings and click **OK**.

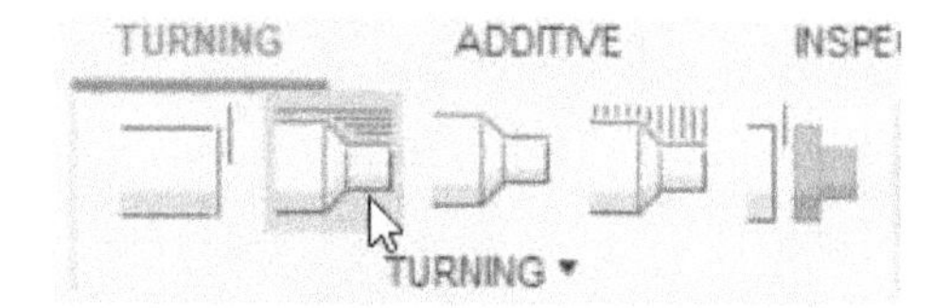

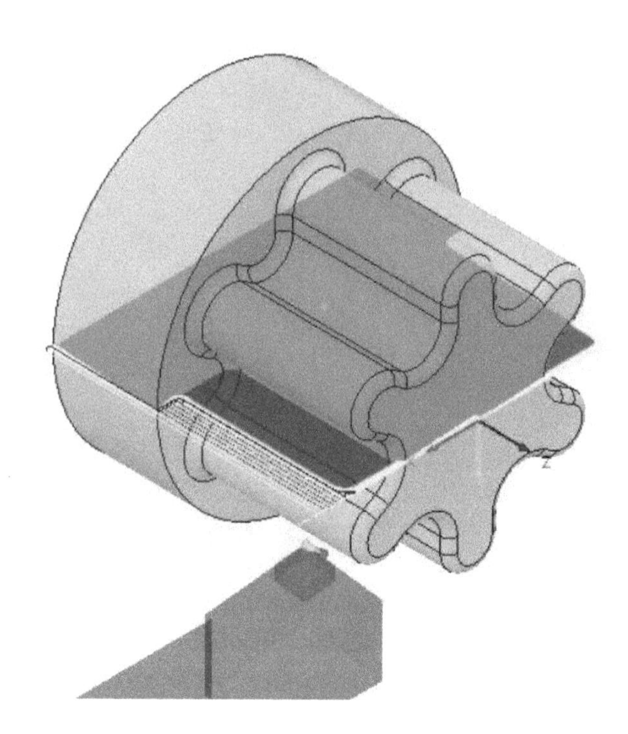

18. On the Toolbar, click **Milling > Multi-axis > Rotary**.
19. On the **Rotary** dialog, click the **Select** button next to the **Tool** option.
20. Click the **Select** button next to the **Tool** option.
21. Right-click on the **Ø3/16" L0.725" (2/16" Flat Endmill)** tool.
22. Select the **Copy tool** option. Next, click **All > Local > Library**.
23. Right-click and select the **Paste tool** option.
24. Select the copied tool, right-click, and then select **Edit Tool**.
25. Type **3/16" Ball Endmill-Copy** in the Description box.
26. Click the **Cutter** tab. Next, enter the following values in the Geometry section.

    Diameter: 0.1875 in
    Shaft diameter: 0.1875 in
    Overall length: 5
    Length below holder: 3
    Shoulder length: 2
    Flute length: 2

27. Click the **Accept** button. Next, click the **Select** button.
28. Click the **Geometry** tab; notice the orange and the green planes. The orange plane represents the front, and the green plane represents the back.

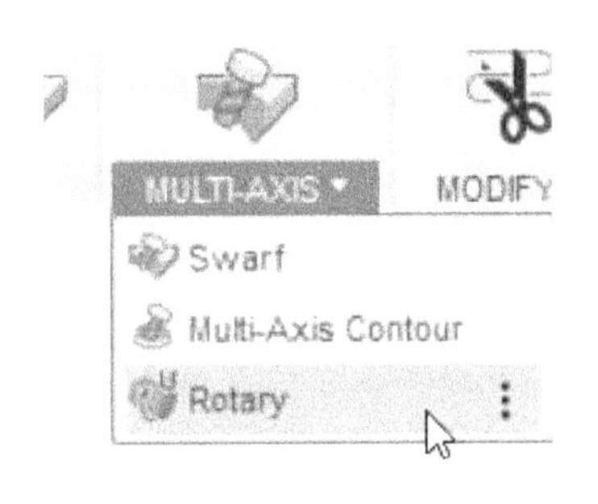

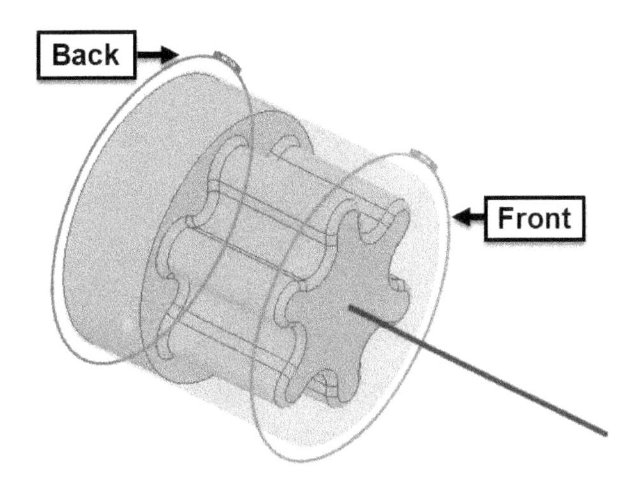

29. Click and drag the green plane towards the front up to a distance of 1.2 inches. Alternatively, you can enter 1.2 in the **Offset** box available in the **Back** section.

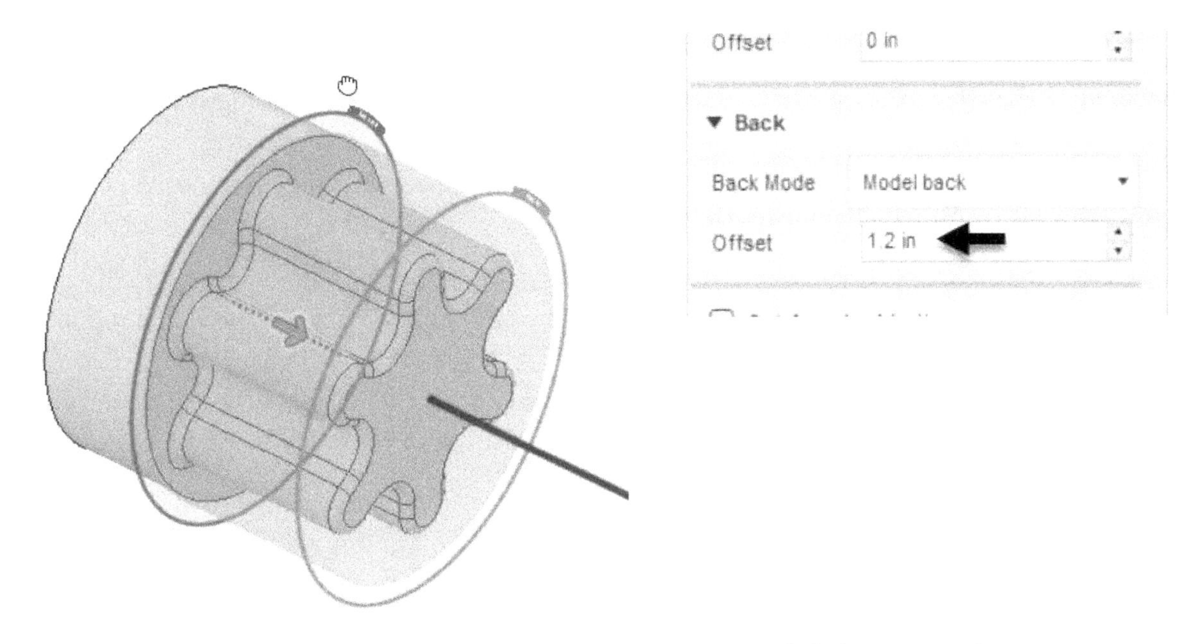

30. Type **0.1** in the **Offset** box available in the **Front** section. Next, click the **Passes** tab.
31. Under the **Rotary Passes** section, select **Style > Spiral**.
32. Right-click in the **Stepover** box, and then select **Edit Expression**.
33. Type **tool_diameter * 0.4** in the **Edit Expression** dialog and click **OK**.
34. Click **OK**.

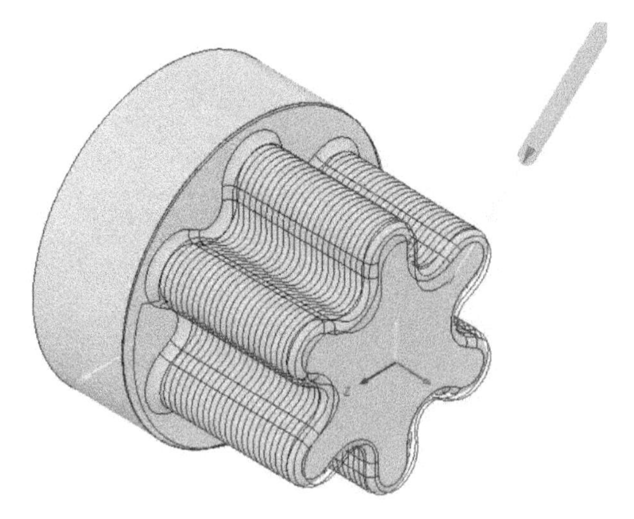

Notice that the toolpath is generated in a spiral fashion. There is only one Lead-in and Lead-out.

35. In the Browser, right-click on the **Rotary** toolpath and select **Duplicate**.
36. Right-click on the duplicated toolpath and select **Edit**.
37. Click the **Passes** tab.
38. Under the **Rotary Passes** section, select **Style > Line**. Note that the **Stepover** units are changed to degrees.
39. Type **3** in the **Stepover** box.
40. Click the **Linking** tab.
41. Select **Retraction Policy > Minimum retraction** and click **OK**.

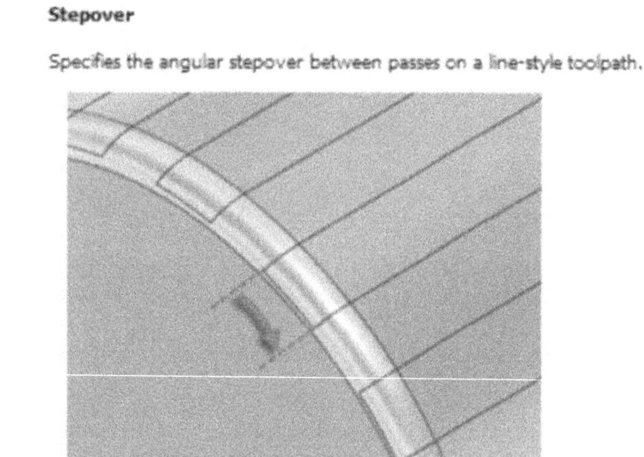

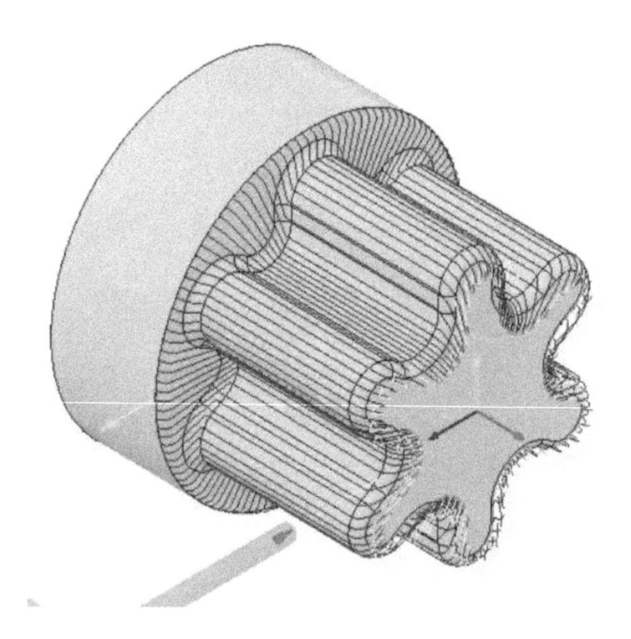

Notice that the passes are generated along the Z-axis.

42. In the Browser, right-click on the second **Rotary** toolpath and select **Duplicate**.
43. Right-click on the new duplicated toolpath and select **Edit**.
44. Click the **Passes** tab. Under the **Rotary Passes** section, select **Style > Circular**.
45. Right-click in the **Stepover** box, and then select **Edit Expression**.

46. Type **tool_diameter * 0.2** in the **Edit Expression** dialog and click **OK**.
47. Click **OK**.

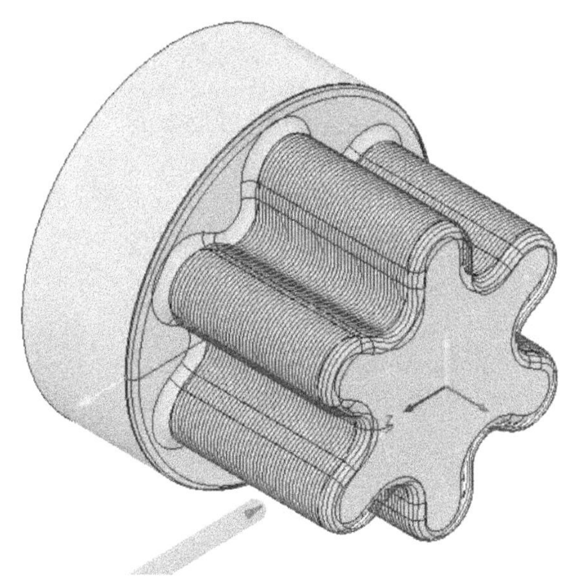

48. Simulate all the toolpaths.
49. Save and close the design file.

Tutorial 24

Creating the Hole Recognition Operation

The **Hole Recognition** operation is used to recognize all holes automatically and generate toolpaths.

1. Download the **Tutorial 24** file, and then upload it to Fusion 360. Next, open the uploaded file.
2. On the Toolbar, click **Milling > Drilling > Hole Recognition**.

   The **Hole Recognition** dialog pops up with all the recognized holes.

3. Select the 0.5-inch hole from the **Hole Recognition** dialog and notice that the 0.5 in holes are highlighted in the geometry. Also, notice that the **Spotdrill & Drill Hole** operation is specified as the **Action**.

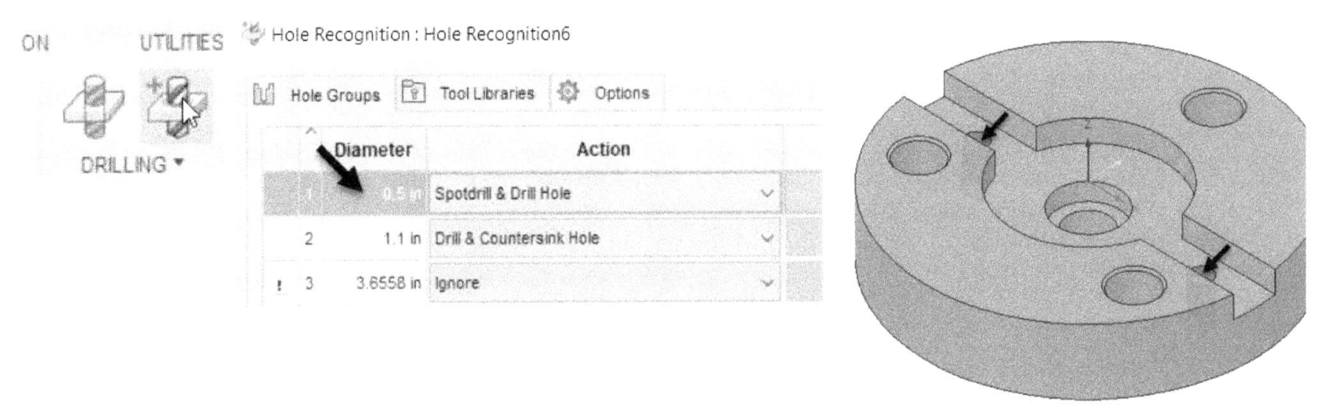

4. Select the **3.65578 in** diameter hole and notice that the pocket features above the counterbored hole are included.

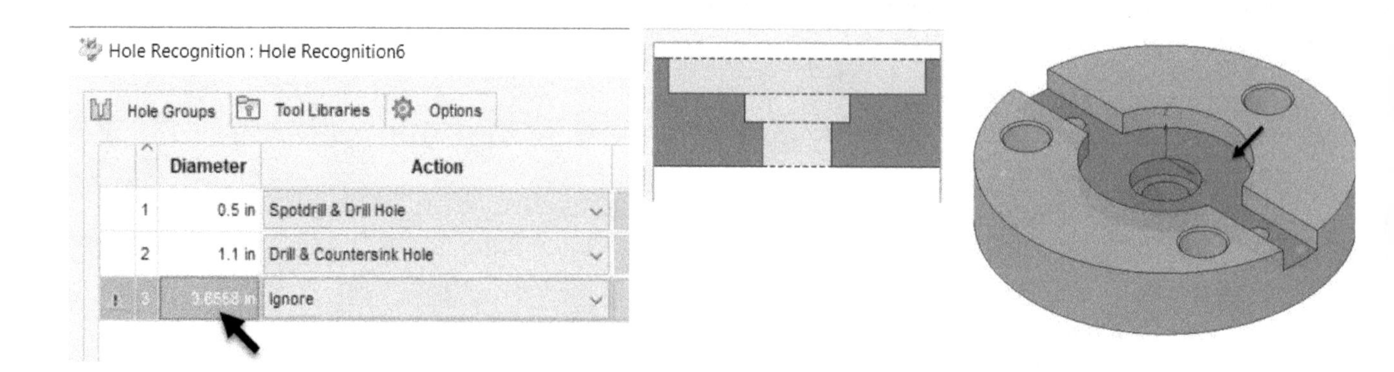

5. Click the **Delete top segment** icon located at the bottom of the **Hole Recognition** dialog.
6. Select **Spot drill, Drill, Counterbore Hole** from the **Actions** drop-down next to the **1.5 in Diameter** hole.

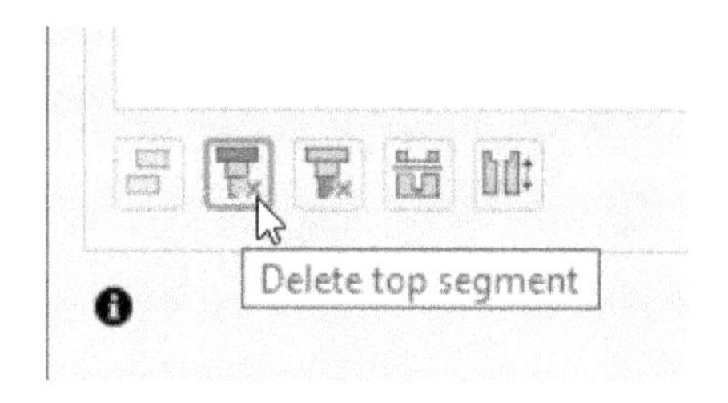

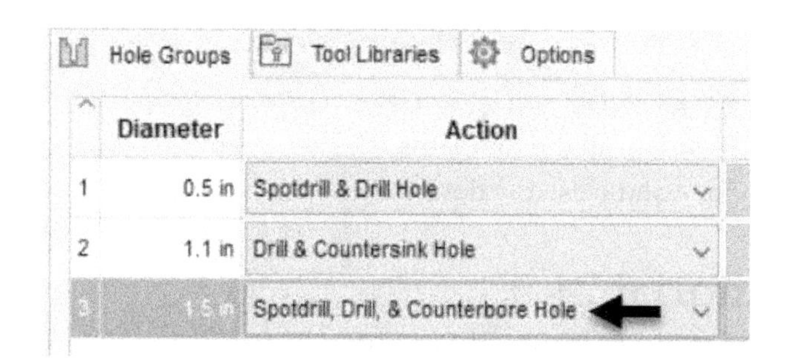

7. Click the **Tool Libraries** tab.
8. Check the **Sample Tools - Inch** and **Tutorial - Inch** options. Next, uncheck the remaining options.
9. Click the **Options** tab, and then select **Organize Operations > Minimize tool changes**.
10. Click **OK**; the toolpath is generated.

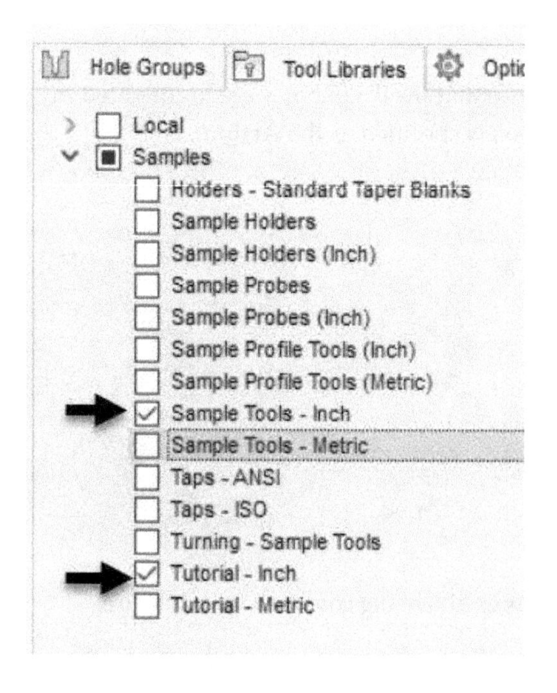

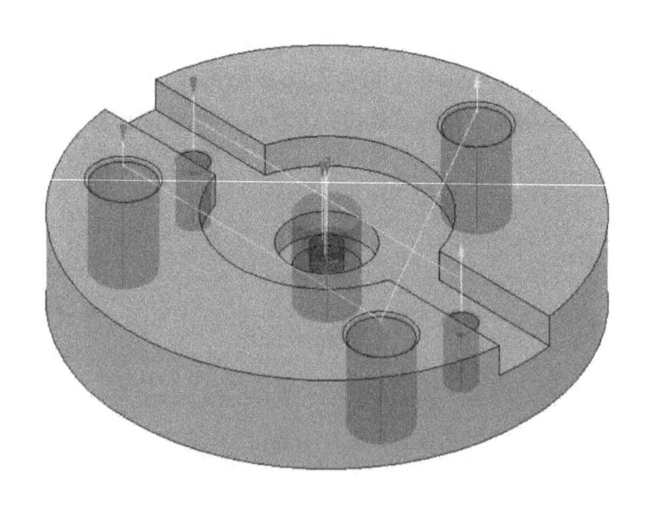

Notice that the Countersink operation is not completed as there no tool available of the required size.

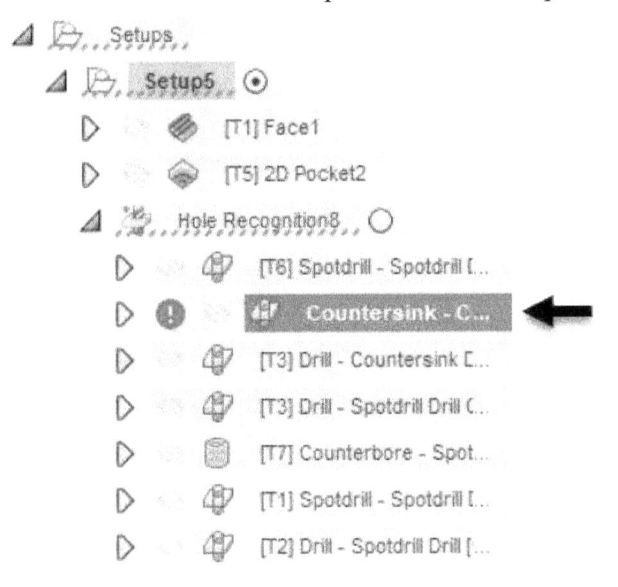

11. Right-click on the countersink operation, and then select **Edit**.
12. Click the **Select** button next to the **Tool** option. Next, click **All > Local > Library**.
13. Click the **New tool** icon. Next, select **Hole making > countersink**.
14. Type **counter sink** in the **Description** box.
15. Click the **Cutter** tab and specify the following values.

    Number of flutes: 1

    Material: HSS

    Diameter: 1.25 in

    Shaft Diameter: 0.5 in

    Tip angle: 82 degrees

    Overall Length: 2.75 in

    Length below holder: 2 in

    Shoulder length: 1.18 in

    Flute Length: 1 in

16. Click **Accept** and **Select**. Next, click **OK** on the **Hole** dialog.
17. Right-click on the **Hole recognition** operation and select **Simulate**.
18. Simulate the operations and close the **Simulate** dialog.

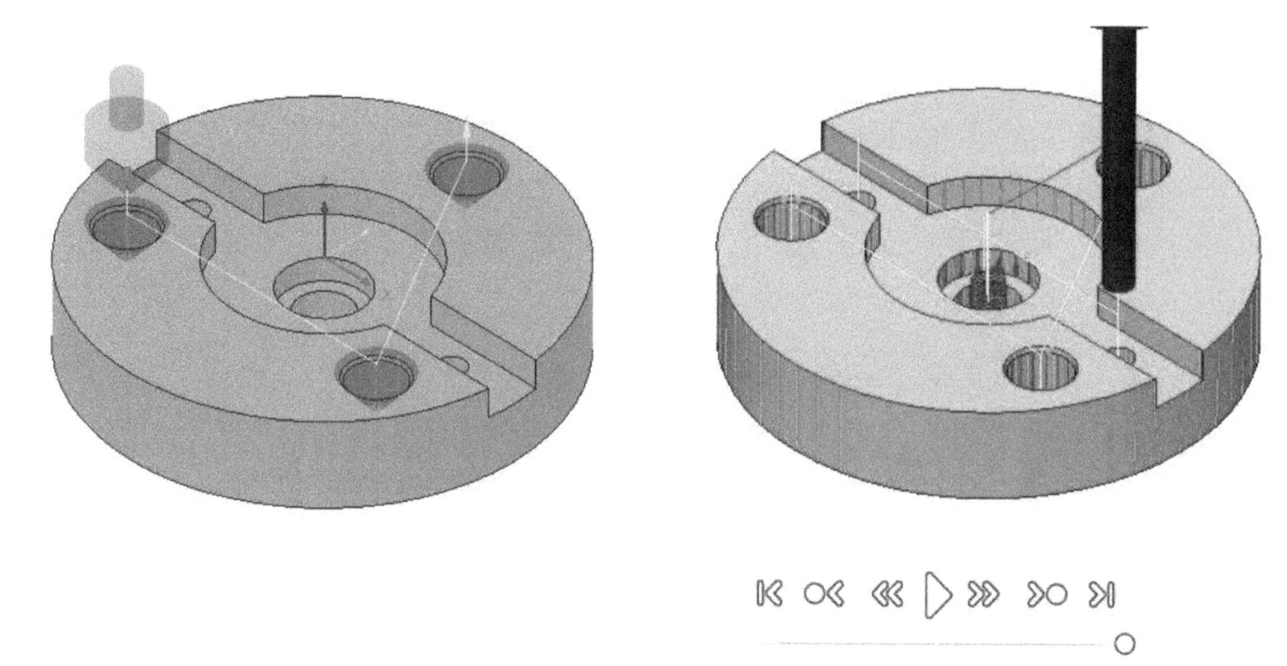

19. Save and close the design file.

Tutorial 25

Creating the Setup for Turning Operations

In this section, you create the Work Coordinate System, Machine Setup, and Stock for the following part.

1. Download the **Tutorial 25** file, and then upload it to Fusion 360. Next, open the uploaded file.
2. On the Toolbar, click **Turning > Setup > Setup**.
3. On the **Setup** dialog, click the **Select** button next to the **Machine** option.
4. On the **Machine Library** window, select **Capabilities > Turning** from the **Filters** tab.
5. Select **Autodesk Generic Mill-Turn Lathe**.

However, you can import your machine configuration by selecting **My Machines > Local** from the tree and then click the **Import** icon.

6. Click the **Select** button.
7. On the **Setup** dialog, select **Operation Type > Turning or mill/turn**.
8. Under the **Work Coordinate System (WCS)** section, select **Orientation > Select Z axis/plane and X axis**.
9. Select the cylindrical face to define the Z axis.
10. Select the axis pointing upwards to define the X-axis.
11. Check the **Flip Z Axis** option if the Z-axis is pointing towards the left.
12. Check the **Flip X Axis** option if the X-axis is pointing downwards.
13. Select **Origin > Stock front**.

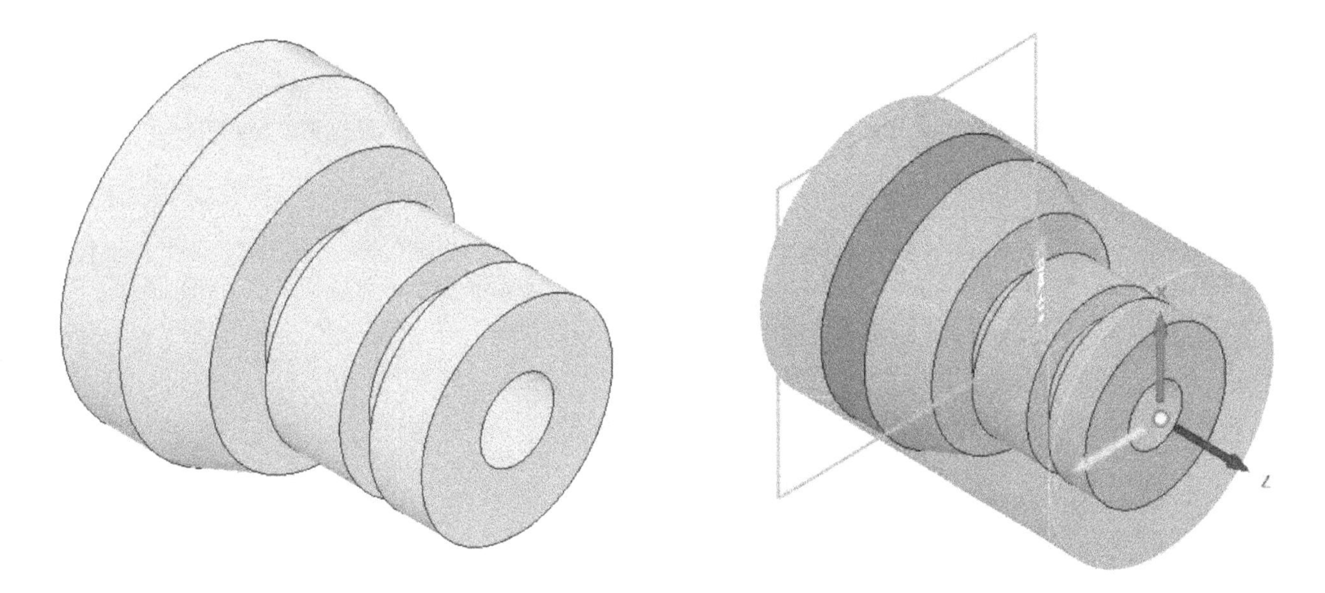

The Work Coordinate System that you defined is used for the turning operations. The turning tool toolpaths are placed in the ZX-plane. All the machining contours are placed on the XZ-plane of the defined Work Coordinate System, as shown.

14. Click the **Stock** tab. Next, select **Mode > Relative size cylinder**.
15. Type **0.08** in the **Radial Stock Offset, Frontside Stock Offset**, and **Round Up to Nearest** boxes.
16. Type **1** in the **Backside Stock Offset** box.
17. Click the **Post Process** tab. Next, type **1001** in the **Program Name/Number** box.
18. Type **Turning Sample Part** in the **Program Comment** box.
19. Click **OK**.

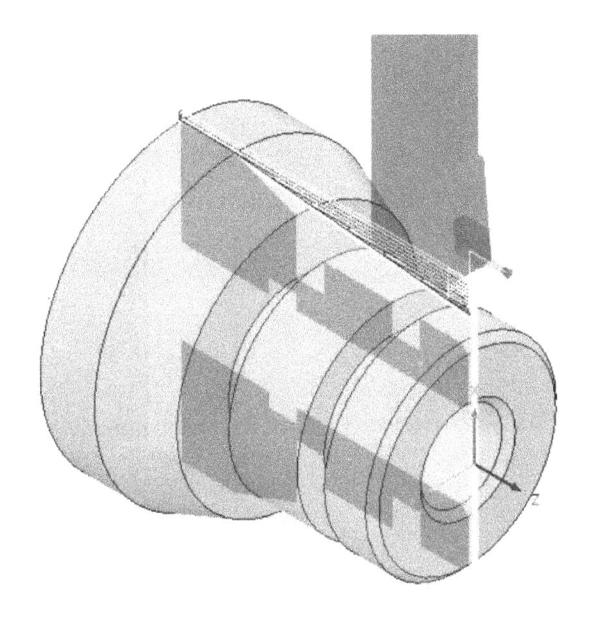

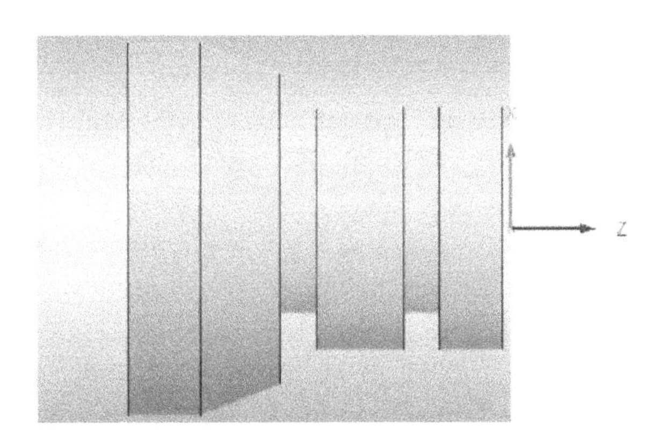

Creating the Turning Face operation

This operation is created at the vertical face at the front of the part. It is used to remove excess stock material from the front portion of the part. The tool moves along the X-axis of the Work Coordinate System.

1. On the Toolbar, click **Turning > Turning > Turning Face**.

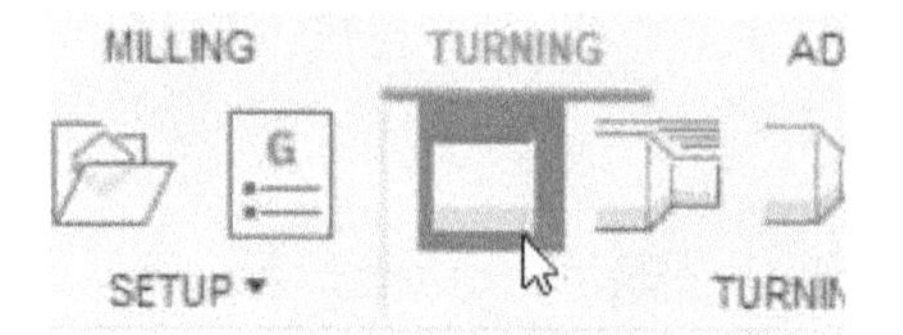

2. On the **Face** dialog, click the **Select** button next to the **Tool** option.
3. On the **Select Tool** window, select the **Turning - Sample Tools** folder from the **Fusion 360 Library**.
4. Select the **CNMT (CNMT Right Hand)** tool from the tool list.
5. Click the **Select** button located at the bottom-right corner.
6. Click the **Geometry** tab and make sure that the **Front Mode** is set to **Model front**. This option enables you to machine the front end face.
7. Click the **Passes** tab and check the **Multiple Passes** option.
8. Check the **Calculate Number of Stepovers** option to calculate the stepover distance automatically.

Check the **Finishing Passes** option if you want to add a finishing pass to the operation.

9. Click the **Linking** tab and select **Retraction Policy > Minimum Retraction**.

Notice that the **Linear Lead-In Length** and **Linear Lead-In Angle** are set to 0.08 and 45 degrees, respectively.

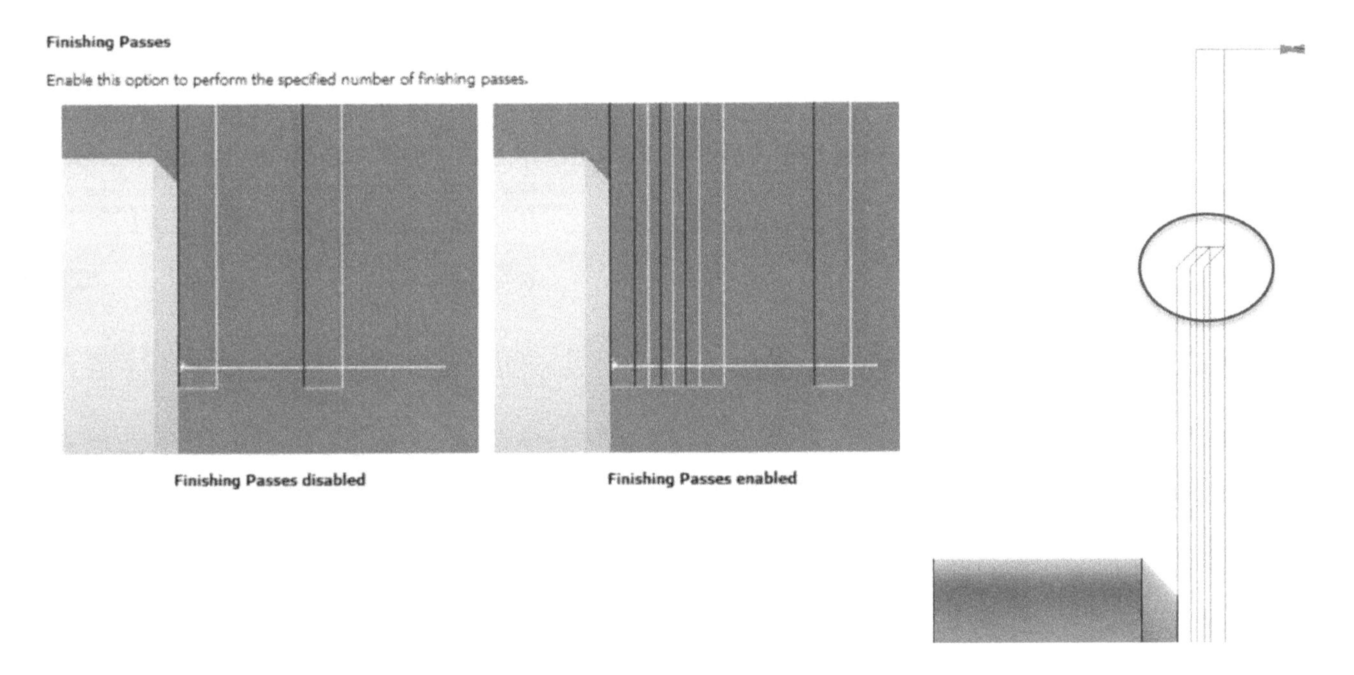

10. Click **OK** on the **Face** dialog.
11. Simulate the **Face** operation and close the **Simulate** dialog.

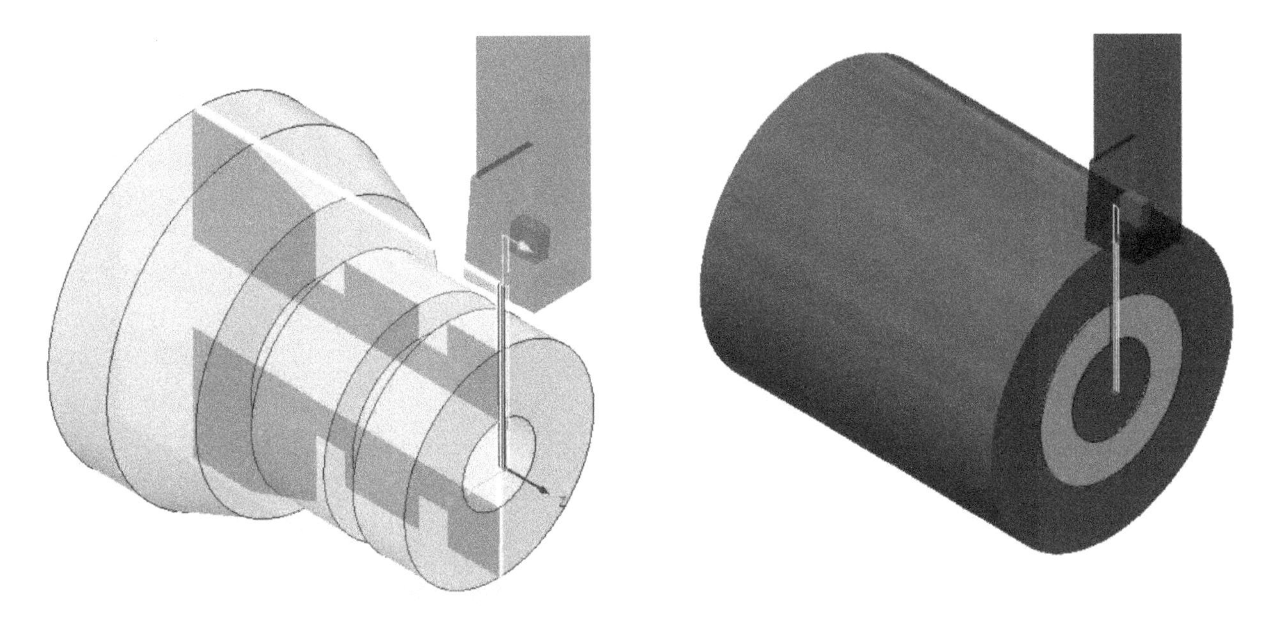

Creating the Turning Profile Roughing operation

This operation is used to remove a large amount of stock material from the part's outer diameter or inner diameter.

1. On the Toolbar, click **Turning > Turning > Turning Profile Roughing**.

   The tool used for the **Face Turning** operation is selected automatically.

2. Under the **Mode** section, select **Turning Mode > Outside profiling**.
3. Click the **Passes** tab and select **Direction > Front to back**.
4. Check the **No Dragging** option under the **Passes** section. This option is useful while machining the vertical portions of the part. It allows the tool to move only in a positive direction. It will avoid any negative pressure on the insert.

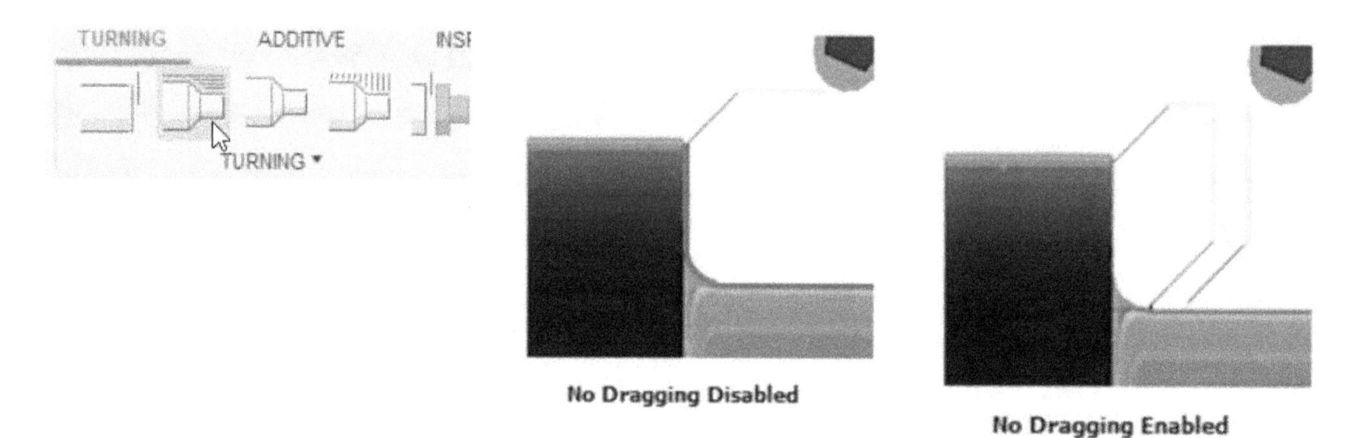

No Dragging Disabled

No Dragging Enabled

5. Under the **Stock to Leave** section, set the **X Stock to Leave** and **Z Stock to Leave** values to **0.015**.

X Stock to Leave

Controls the amount of material to leave in the radial direction.
Used for leaving stock on the ID or OD of the part.

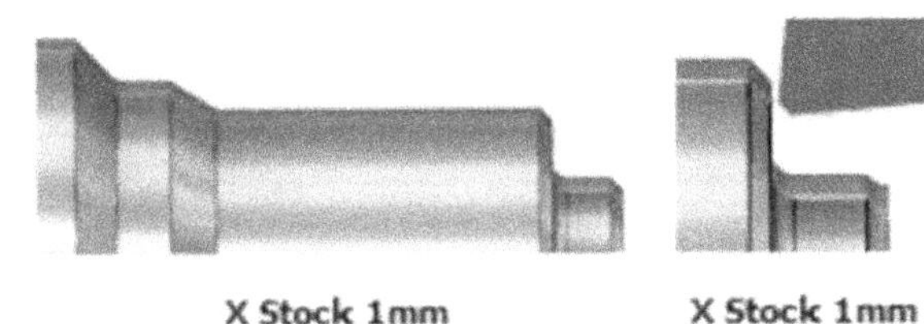

X Stock 1mm X Stock 1mm

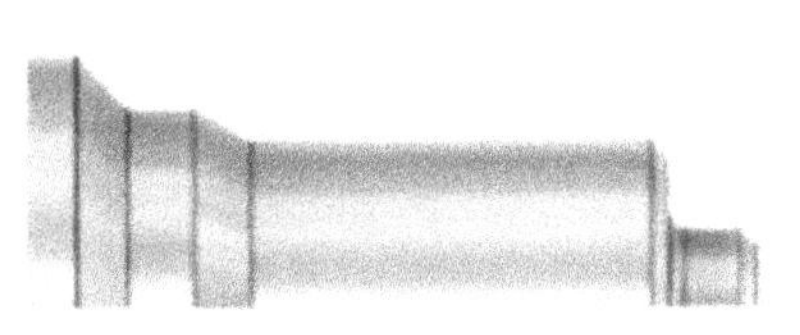

X Stock 0.0

X and Z Stock -1mm

For surfaces that are not exactly horizontal, the program interpolates between the Axial Stock value (wall) and the Radial Stock values. The stock left in the radial direction on these surfaces might be different from the specified value, depending on surface slope and the axial stock to leave value.

Changing the **X Stock to Leave** value automatically sets the **Z Stock to Leave** to the same amount. You can manually enter a different Axial Stock value.

If you plan on roughing and finishing with different tools, set the rough operation to the amount of stock to leave for the next finishing operation.

Negative Stock to Leave You may use a negative value. This will cause the tool to machine material beyond the edge of your model shape.

NOTE: - Maximum negative stock to leave must be less than the tool nose radius.
With an insert nose radius of .032, the largest negative stock would be -.032
You can not compensate past the theoretical tip of the tool.

6. Leave the default settings and click **OK**.
7. Simulate the **Profile Roughing** Operation and close the **Simulate** dialog.

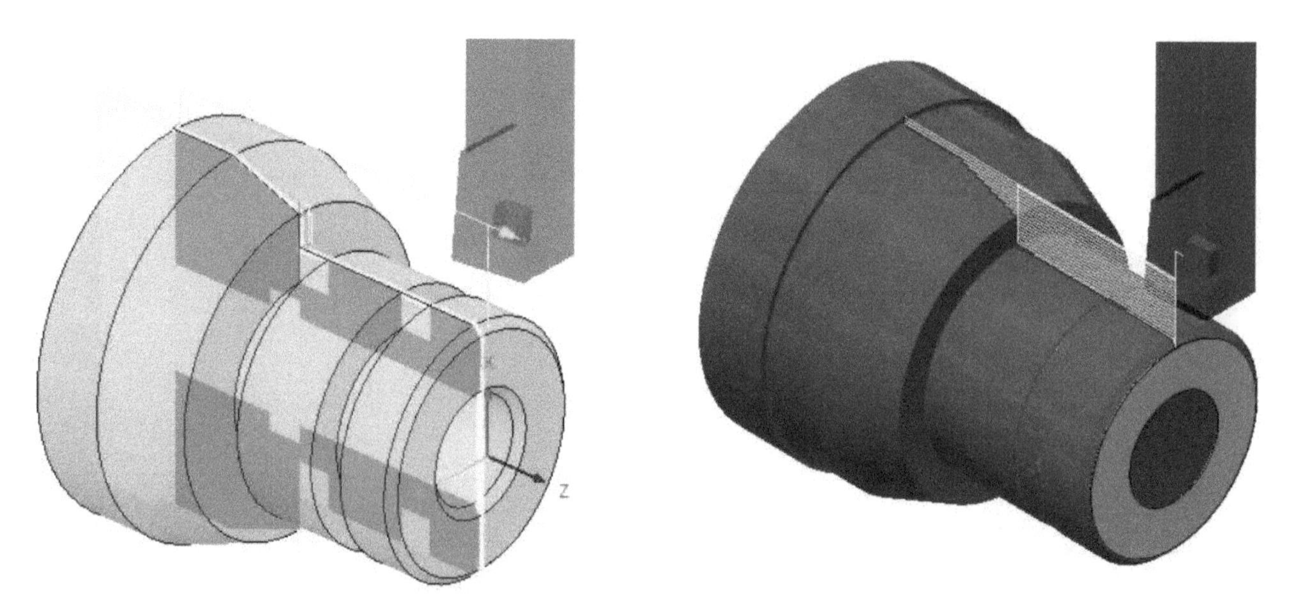

Select the **Profile Roughing** operation from the Browser, and then change the view orientation to front. Notice that the toolpath ends exactly at the backside of the part with no clearance for a cut-off tool. Also, there are slight dip-ins at the grooves.

8. Right-click on the **Profile Roughing** operation and select **Edit**.
9. Click the **Geometry** tab and type **-0.2** in the **Offset** box available in the **Back** section.
10. Click the **Passes** tab and select **Grooving > Don't allow grooving**. Next, click **OK**.

Notice that the toolpath is extended beyond the back face of the part. Also, the dip-ins disappear from the groove locations.

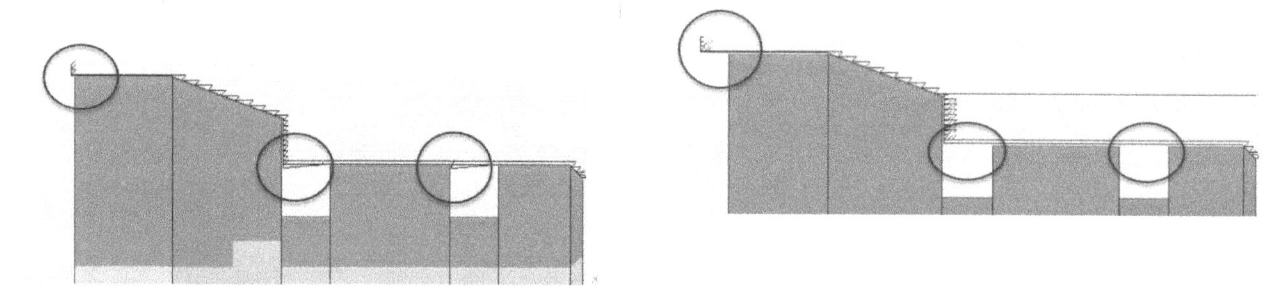

Creating the Turning Profile Finishing operation

After creating the profile roughing operation, you need to create the finishing operation.

1. On the Toolbar, click **Turning > Turning > Turning Profile Finishing**.
2. Click the **Geometry** tab and check the **Rest Machining** option.
3. Select **Source > From previous operation(s)**.
4. Click the **Passes** tab and select **Grooving > Don't allow grooving**.
5. Leave the **Number of Stepovers** value to 1.
6. Check the **No Dragging** option.
7. Click **OK**.

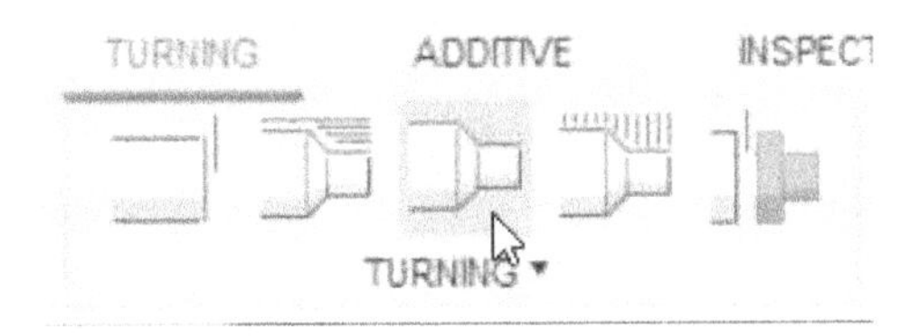

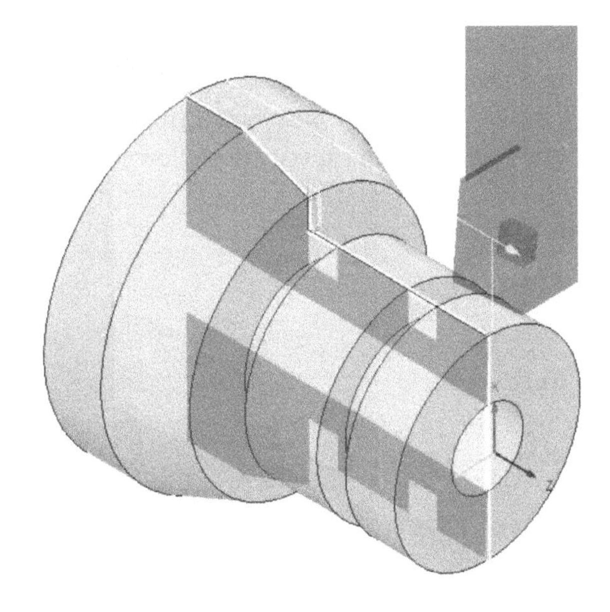

Creating the Turning Groove operation

A groove is a feature that is closed on both sides and below the surrounding geometry's surface. The **Turning Groove** operation creates external or internal grooves either on the radial geometry or an axial geometry (wall).

1. On the Toolbar, click **Turning > Turning > Turning Groove**.
2. Click the **Select** button next to the **Tool** option.
3. On the **Select Tool** window, select the **Turning – Sample Tools** folder from the **Fusion 360 Library**.
4. Select the **OD Grooving** tool from the tool list.
5. Click the **Select** button located at the bottom-right corner.
6. Click the **Geometry** tab. Next, check the **Rest Machining** option.
7. Select **Source > From previous operation(s)**.
8. Click the **Passes** tab. Make sure that the **Finishing Passes** option is checked.
9. Type 0.01 in the **Stepover** box. Next, click **OK**.

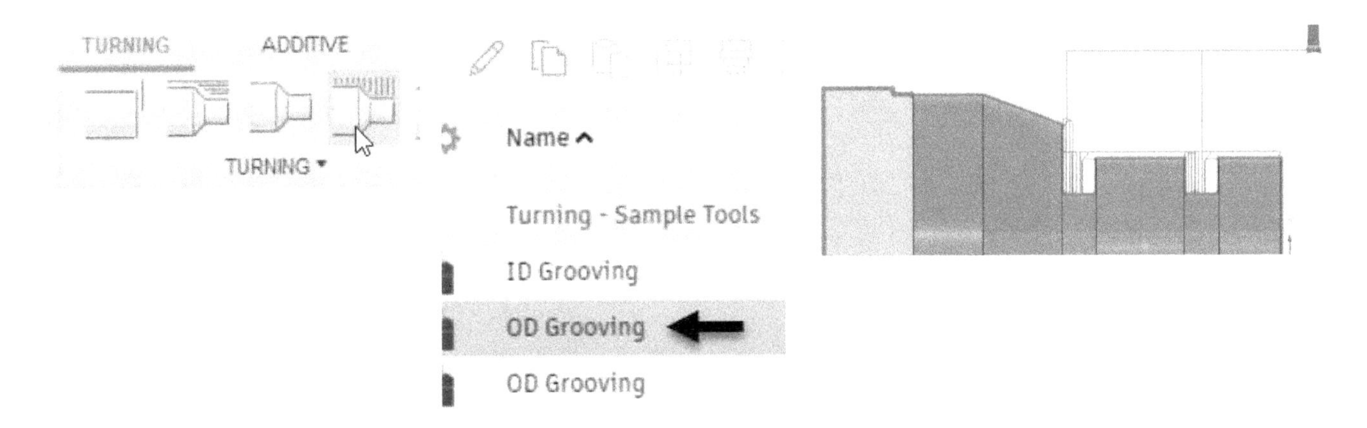

Creating Drilling operations

These operations help you to drill along the rotation axis.

1. On the Toolbar, click **Turning > Drilling > Drill**. Next, click the **Select** button next to the **Tool** option.
2. On the **Select Tool** window, select the **Local > Library** folder.
3. Click the **New Tool** icon. Next, select **Hole Making > Spot drill**.
4. Click the **Cutter** tab and specify the following values.

Diameter: 1.125 in

Shaft diameter: 1.125 in

Overall length: 5.625 in

Length below holder: 5.625 in

Shoulder length: 3.375 in

Flute length: 2.25 in

Tip diameter: 0

Tip angle: 118 degrees

5. Click the **Cutting Data** section. Next, enter the following values.

Spindle Speed: 1188 rpm

Cutting feedrate: 16.63 in/min

Feed per revolution: 0.014 in

6. Click **Accept** and **Select**.
7. Click the **Geometry** tab and select the cylindrical face of the hole.
8. Click the **Heights** tab. Under the **Bottom Height** section, select **From > To chamfer width**.
9. Type **0.01** in the **Offset** box, and click **OK**.

10. On the Toolbar, click **Turning > Drilling > Drill**. Next, click the **Select** button next to the **Tool** option.
11. On the **Select Tool** window, select the **Fusion 360 Library > Sample Tools - Inch** folder.
12. Select **Hole Making** from the **Filter** tab located at the top-right corner. Next, select **Type > drill**.
13. Select the **63/64" L9.94375 (63/64)** tool from the tool list.
14. Select **Aluminum – Drilling** from the **Cutting Data** section. Next, click the **Select** button.
15. Click the **Geometry** tab. Next, select the cylindrical face of the hole.
16. Click the **Heights** tab. Next, under the **Bottom Height** section, select **From > Model bottom**.
17. Click the **Passes** tab. Next, select **Cycle Type > Chip breaking – partial retract**.
18. Type **0.1** in the **Pecking Depth** box. Next, click **OK**.

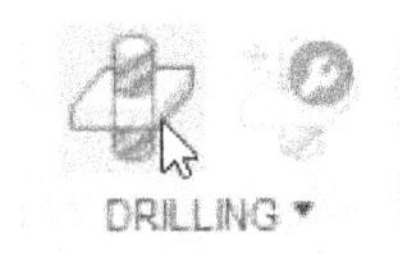

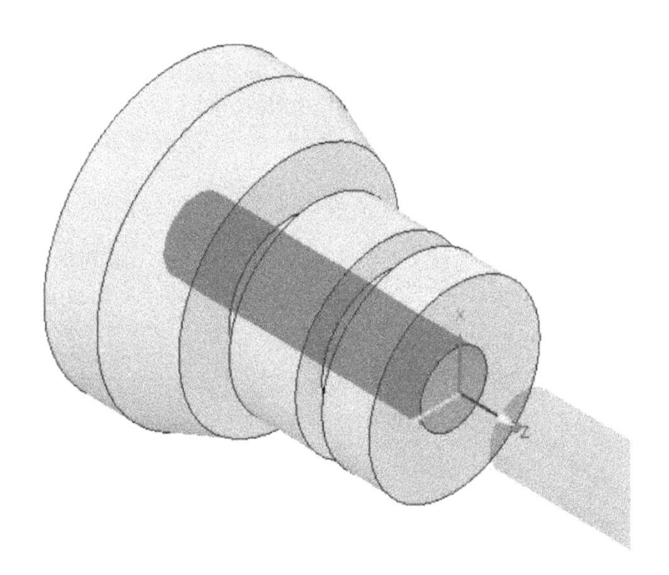

Next, you need to ream the hole.

19. On the Toolbar, click **Turning > Drilling > Drill**. Next, click the **Select** button next to the **Tool** option.
20. On the **Select Tool** window, select the **Local > Library** folder. Next, click the **New Tool** icon.
21. Select **Hole Making > reamer**. Next, click the **Cutter** tab and specify the following values.

Diameter: 1 in

Shaft diameter: 1 in

Overall length: 7 in

Length below holder: 7 in

Shoulder length: 5

Flute length: 4

22. Click the **Cutting Data** section. Next, enter the following values.

Spindle Speed: 955 rpm

Surface Speed: 250 ft/min

Plunge feedrate: 15.28 in/min

Feed per revolution: 0.016 in

Retract feedrate: 15.28 in/min

23. Click **Accept** and **Select**.
24. Click the **Geometry** tab and select the cylindrical face of the hole.
25. Click the **Heights** tab.
26. Under the **Bottom Height** section, select **From > Model Bottom**.
27. Click the **Cycle** tab and select **Cycle Type > Reaming feed out**.
28. Click **OK**.

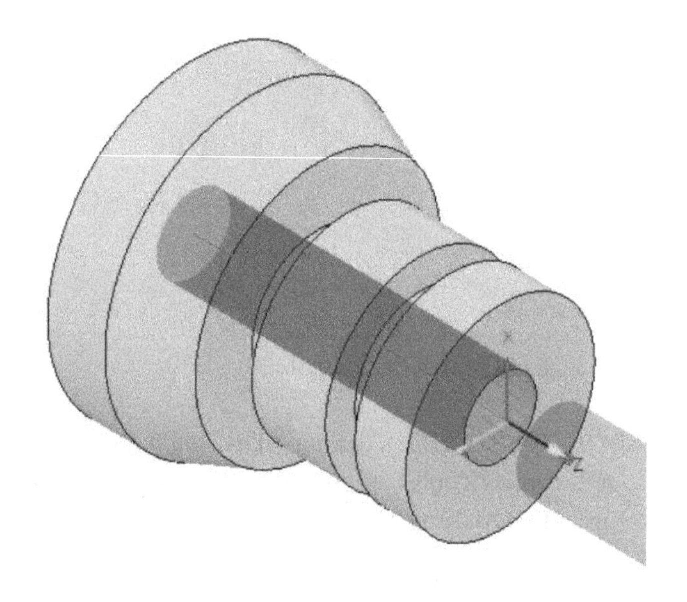

Creating the Thread Turning operation

This operation helps you to perform threading along the rotation axis.

1. On the Toolbar, click **Turning > Turning > Turning Thread.**

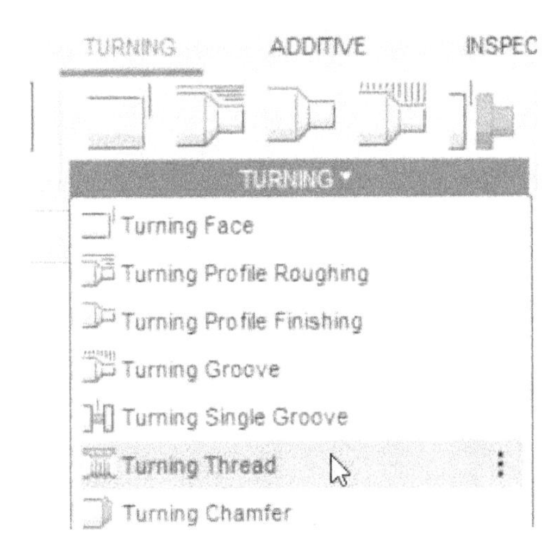

2. Click the **Select** button next to the **Tool** option.
3. On the **Select Tool** window, select the **Fusion 360 Library > Turning - Sample Tools** folder.
4. Select the **OD- Threading** tool from the tool list.
5. Click the **Select** button. Next, select **Turning Mode > Outside Threading.**
6. Click the **Geometry** tab.
7. Select the cylindrical face of the model, as shown.

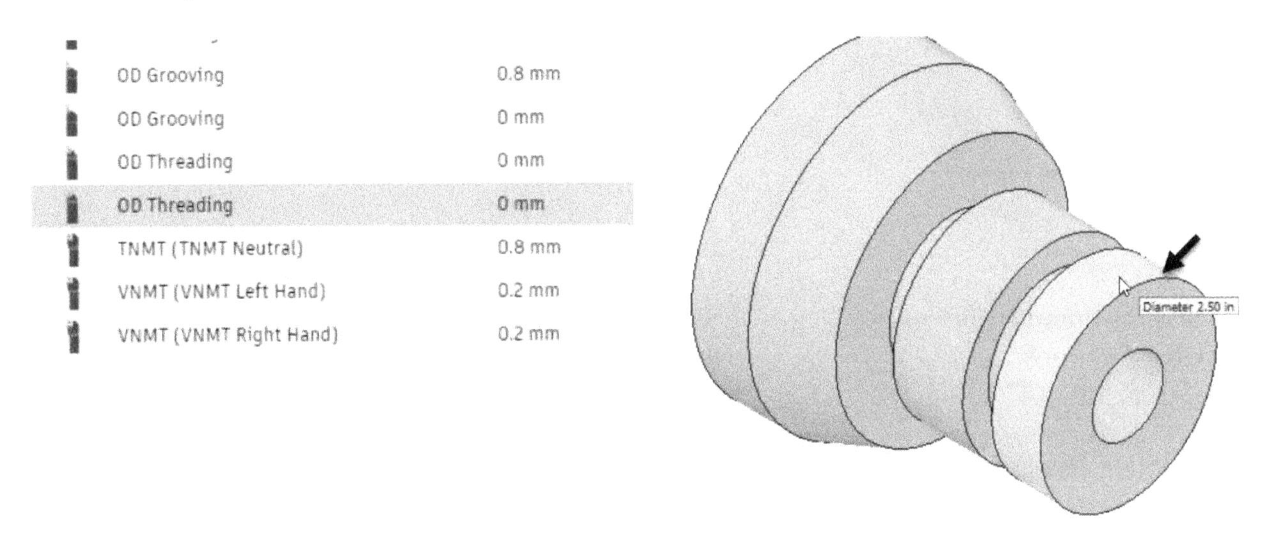

8. Type 0.2 in the **Backside Stock Offset** box.
9. Click the **Radii** tab and select **From > Model OD.**
10. Type **0.12** in the **Offset** box.
11. Click the **Passes** tab.
12. Select **Threading Hand > Right Handed.**
13. Type **0.714** in the **Thread Pitch** box.

14. Select **Infeed Mode > Reduced infeed.**

Infeed Mode

The infeed is the depth of cut per pass and is critical in threading. Each successive pass engages a larger portion of the cutting edge of the insert. There are three infeed mode options.

Constant infeed - With this option, the cutting force and metal removal rate can increase dramatically from one pass to the next. **Example:** When producing a 60 degree thread form using a constant 0.010 inch infeed per pass, the second pass removes three times the amount of metal as the first pass. The amount of metal removed continues to grow exponentially with each subsequent pass.

Reduced infeed - This option maintains more realistic cutting forces and reduces the depth of cut with each pass. Because it avoids the increased metal removal rate of constant infeed, it is the recommended setting.

Alternate infeed - Similar to the Reduced infeed, but this method alternates between feeding in from the left and right sides of the thread profile.

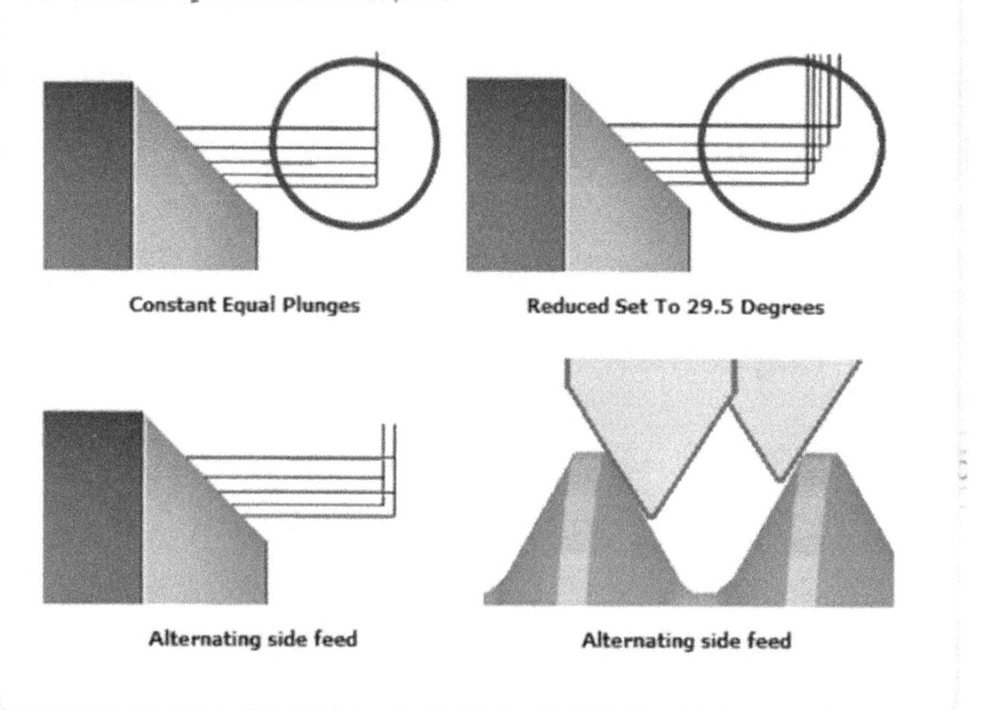

15. Type **30** in the **Infeed angle** box.
16. Click **OK**.

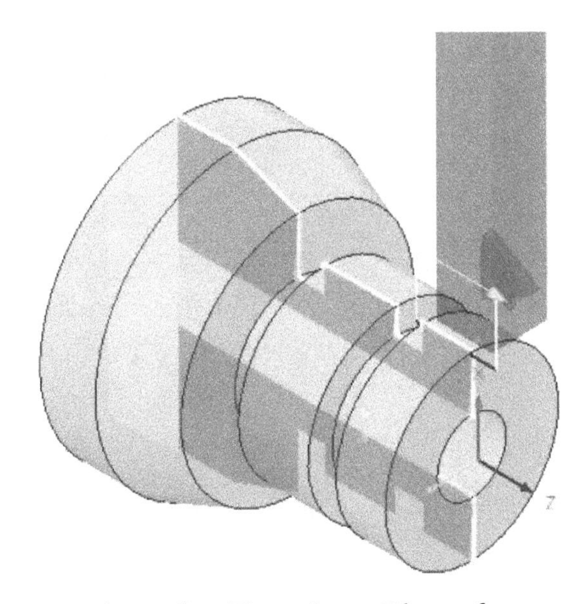

Creating the Turning Chamfer operation

This operation helps you to create chamfers.

1. On the Toolbar, click **Turning > Turning > Turning Chamfer**.

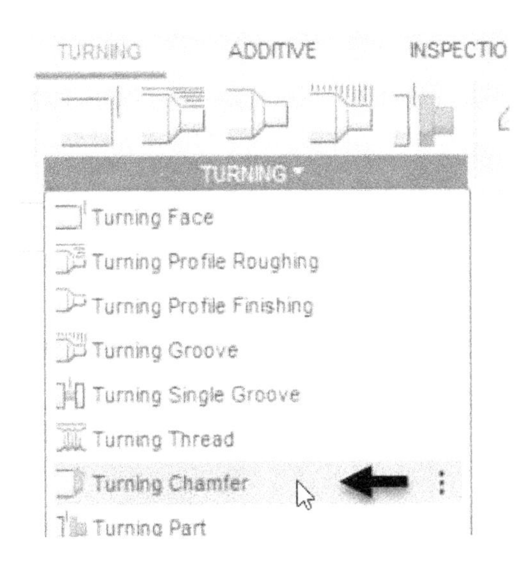

2. Click the **Select** button next to the **Tool** option.
3. On the **Select Tool** window, select the **Fusion 360 Library > Turning - Sample Tools** folder.
4. Select the **CNMT (CNMT Right Hand)** tool from the tool list.
5. Click the **Select** button located at the bottom-right corner.
6. Select **Turning Mode > Outside chamfer**.
7. Click the **Geometry** tab.
8. Select the circular edge of the model, as shown.
9. Click the **Passes** tab. Next, check the **Multiple Passes** option.
10. Type **0.05** in the **Stepover** box.
11. Type **0.1** in the **Chamfer Width** box.

12. Type **0.04** in the **Chamfer Extension** box.
13. Type **45** in the **Chamfer Angle** box. Next, click **OK**.

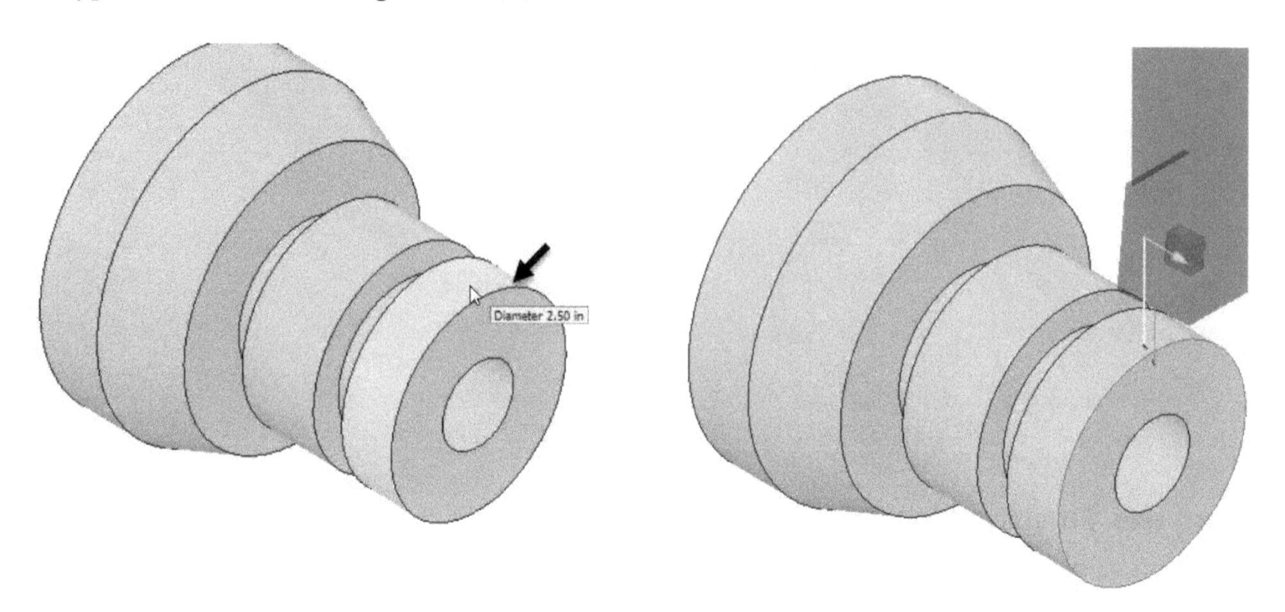

Creating the Turning Part operation

This operation helps you to cut-off the part from the remaining stock at the backside.

1. On the Toolbar, click **Turning > Turning > Turning Part**.

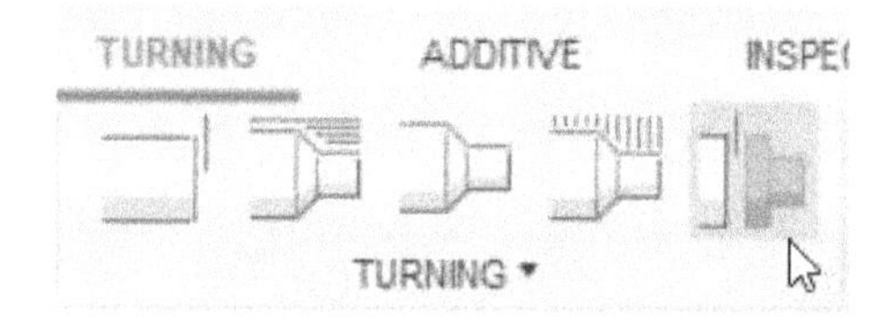

Notice that the grooving tool is selected automatically. Also, the toolpath is generated at the backside of the part.

2. Click the **Geometry** tab. Next, check the **Edge Break** option.
3. Select **Edge Break Type > Chamfer**. Next, type **0.1** in the **Chamfer Width** box.
4. Leave the **Chamfer Angle** value to **45** degrees.
5. Click the Passes tab and check the **Use Pecking** option. It will create multiple steps up to the final depth. This option is useful if the material produces long chips. Also, it allows the coolant the reach the tool-tip.
6. Click **OK**.

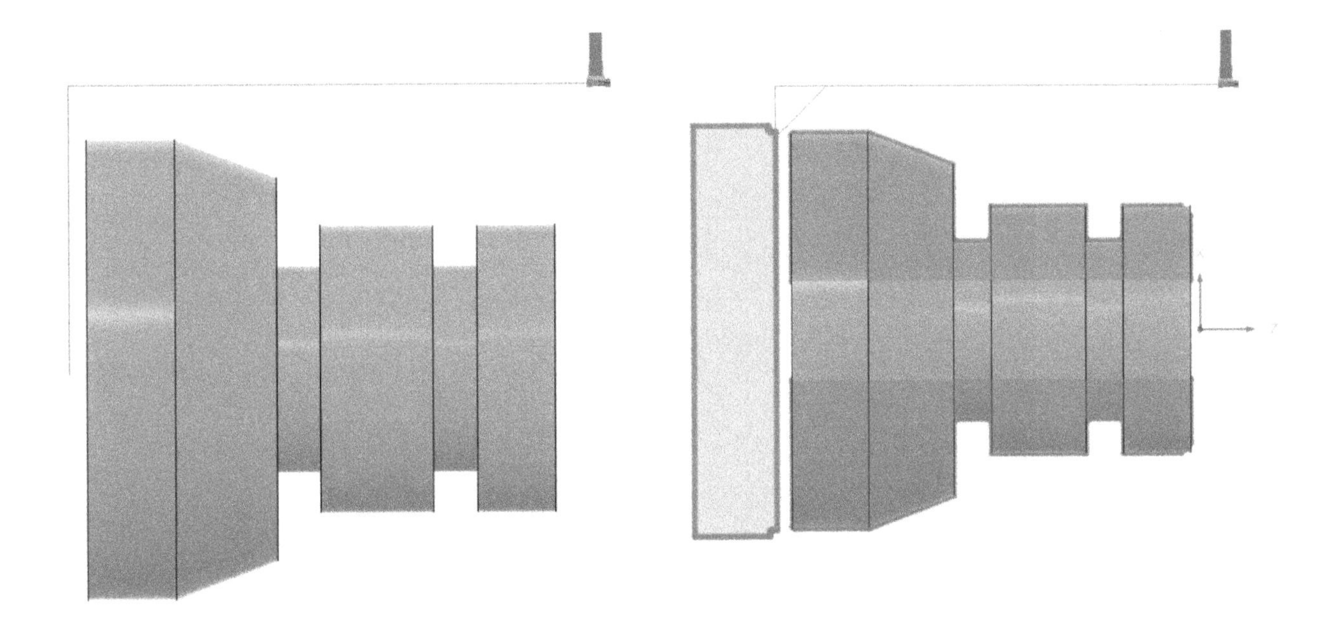

7. Save and close the file.

Index

Printed in the USA
CPSIA information can be obtained
at www.ICGtesting.com
LVHW070749081223
765756LV00013B/435